Springer Theses

Recognizing Outstanding Ph.D. Research

More information about this series at http://www.springer.com/series/8790

Aims and Scope

The series "Springer Theses" brings together a selection of the very best Ph.D. theses from around the world and across the physical sciences. Nominated and endorsed by two recognized specialists, each published volume has been selected for its scientific excellence and the high impact of its contents for the pertinent field of research. For greater accessibility to non-specialists, the published versions include an extended introduction, as well as a foreword by the student's supervisor explaining the special relevance of the work for the field. As a whole, the series will provide a valuable resource both for newcomers to the research fields described, and for other scientists seeking detailed background information on special questions. Finally, it provides an accredited documentation of the valuable contributions made by today's younger generation of scientists.

Theses are accepted into the series by invited nomination only and must fulfill all of the following criteria

- They must be written in good English.
- The topic should fall within the confines of Chemistry, Physics, Earth Sciences, Engineering and related interdisciplinary fields such as Materials, Nanoscience, Chemical Engineering, Complex Systems and Biophysics.
- The work reported in the thesis must represent a significant scientific advance.
- If the thesis includes previously published material, permission to reproduce this must be gained from the respective copyright holder.
- They must have been examined and passed during the 12 months prior to nomination.
- Each thesis should include a foreword by the supervisor outlining the significance of its content.
- The theses should have a clearly defined structure including an introduction accessible to scientists not expert in that particular field.

Arnab De

Ubiquitin Chains: Degradation and Beyond

Doctoral Thesis accepted by
Columbia University
New York, NY

Springer

Author
Dr. Arnab De
Department of Microbiology and Immunology
Columbia University Medical Center
New York, NY, USA

Supervisor
Prof. Sankar Ghosh
Chair, Silverstein and Hutt Family Professor of Microbiology and Immunology
Department of Microbiology and Immunology
Columbia University Medical Center
New York, NY, USA

ISSN 2190-5053 ISSN 2190-5061 (electronic)
Springer Theses
ISBN 978-3-319-14964-6 ISBN 978-3-319-14965-3 (eBook)
DOI 10.1007/978-3-319-14965-3

Library of Congress Control Number: 2015930512

Springer Cham Heidelberg New York Dordrecht London

Printed on acid-free paper

Springer International Publishing AG Switzerland is part of Springer Science+Business Media (www.springer.com)

Dedicated to
The Inspiration of Sachin Tendulkar
&
My Alma Mater
(Presidency College, Calcutta, India
and Columbia University, New York, USA:
two temples of learning, in corners
of the world)

While dedication of a doctoral thesis to places of learning is considered common practice, many have pondered, why dedicate a Ph.D. thesis to a sportsman? A brief word might be apt here.

Sachin Tendulkar is the greatest batsman the cricket world has seen in the post war era. Some of the greatest feats of Sachin Tendulkar (including the first double century ever in history) happened when I was at Columbia University, away from home studying in New York. However, this thesis is not dedicated to the cricketing genius of the man. While he excelled in cricket, many others have done equally well in their respective field of enterprise.

Instead, this thesis is dedicated to Sachin Tendulkar for the way he has inspired the Indian youth of my generation (and me) for the past 25 years. He did so with his immense dedication and love for his work (game of cricket); and his unwavering intent to continue to be better at it, even when he was the best in the world. The confidence of the Indian youth today is a reflection of the certainty that Sachin gradually brought to his cricket; that is his ultimate gift to India.

In my opinion, Tendulkar is an idea, an idea that will be cherished by everyone who believes in work and letting "karma" talk.

Supervisor's Foreword

K48-linked ubiquitination is well known as a signal for degradation. However, there has been a growing body of work which suggests that K63-linked polyubiquitin chains are non-degradative and play a role in regulating cell-signaling. The case for regulatory ubiquitination has been to a large extent predicated upon the presumed deubiquitinase activity of A20, long considered a key regulator of inflammatory responses as mice lacking A20 die from multiorgan inflammation and cachexia. Consistent with its role in limiting inflammatory responses through its deubiquitinase activity, A20 has been implicated in a wide variety of autoimmune diseases and cancer. Multiple biochemical and knockdown approaches in cell culture have suggested that A20 functions as a deubiquitinase by disassembling regulatory K63-ubiquitin chains on upstream signaling molecules such as RIP1; thereby downregulating the NF-κB-mediated inflammatory response. This thesis examines the evidence for regulatory ubiquitination by focusing on A20 and reports the creation and characterization of a knock-in mouse that expresses a mutated form of A20 that selectively lacks the deubiquitinase activity. The knock-in mice surprisingly display completely normal NF-κB activation with no accompanying inflammatory phenotype. Therefore, these results clearly demonstrate that the well-characterized role of A20 in limiting inflammatory responses is due to effects other than deubiquitination of K63-ubiquitin chains. Thus, it might be important to revisit the role of K63-linked polyubiquitination cell-signaling.

Chair, Silverstein and Hutt Family Professor
of Microbiology and Immunology
Department of Microbiology and Immunology
Columbia University Medical Center
New York, NY, USA

Sankar Ghosh, Ph.D.

Abstract

K48-linked ubiquitination is well known as a signal for degradation, and the finding was recognized by the Nobel Prize for Chemistry in 2004. Recent work has suggested that K63-linked ubiquitination does not lead to degradation; instead, it has been proposed to be important for cell-signaling. Indeed, there is a growing body of *in vitro* evidence that suggests that non-degradative K63-linked polyubiquitin chains play an important role in activation of transcription factor NF-κB.

While prompt activation of NF-κB is essential for optimal immune response, it is equally important to terminate the response in order to avoid tissue damage and perhaps even death resulting from organ failure. A20 is an essential inhibitor of NF-κB-mediated inflammation as mice lacking A20 die from multiorgan inflammation and cachexia. Multiple biochemical approaches have suggested that A20 functions as a deubiquitinase by disassembling K63-linked regulatory ubiquitin chains from upstream adapter molecules like RIP1. The importance of K63-linked polyubiquitin chains in abetting cell-signaling has to a large extent been predicated on the deubiquitinase activity of A20. To determine the contribution of the deubiquitinase role of A20 in downregulating NF-κB, we generated and characterized a knock-in mouse lacking the deubiquitinase activity of A20. However, we find that these mice surprisingly display normal NF-κB activation and show no signs of inflammation.

Our work clearly demonstrates that the deubiquitinase activity of A20 is dispensable for downregulating NF-κB and the well-characterized role of A20 in limiting inflammatory responses is due to effects other than deubiquitination of K63-ubiquitin chains. Hence, it might be critical to revisit and explore the general role of K63-linked polyubiquitination in cell-signaling.

Acknowledgements

I am profoundly grateful to Professor Sankar Ghosh for his guidance in life in general and in science in particular. Working at close quarters with him has given me an incredible insight into his vision and leadership, something I will always cherish and hope to put to use for the rest of my life. In addition, I also thank Dr. Boris Reizis, Dr. Uttiya Basu, Dr. Chozha Vendan Rathinam, Prof. David Fidock, and Prof. Max Gottesman at Columbia University, Prof. Dongsheng Cai at the Albert Einstein College of Medicine and Dr. Hediye Erdjument-Bromage and Dr. Paul Tempst at the Proteomics and Microchemistry core facility of Memorial Sloan-Kettering Cancer Center.

I would like to thank all my friends and colleagues at Columbia University, both scientific and nonscientific. I especially thank Dr. Teruki Dainichi (currently a faculty at Kyoto University) with whom I had worked closely during this time. I would also like to thank our departmental administrators (Edith Shumansky and Carla Horne) and our lab-manager (Crystal Bussey) for always being around. Indeed, it is a privilege to receive a Ph.D. from the Department of Microbiology and Immunology at Columbia University.

Finally, I thank my family: my father (Dr. Arun Kumar De), my mother (Mrs. Manjulika De), my little brother (Dr. Arka De), and my lovely wife (Dr. Rituparna De Bose). For the sake of brevity, I would only say "Thank You." One of the better decisions I made during my Ph.D. was to get married to the most understanding and loving person I have known; I know that this is already turning out to be one of the best decisions in life.

I would like to end this part by thanking and paying my tributes to the majestic city of New York, a great place I called home during this time. It will always be a part of me.

Contents

About the Author

Dr. Arnab De obtained his Ph.D. from the Department of Microbiology and Immunology at Columbia University. During his Ph.D., he developed transgenic mice to study the mechanism of action of a critical tumor-suppressor called A20. He is also well known for having developed peptide-based prodrugs as therapeutics for diabetes. His work has been reported by the media and has resulted in multiple patents and publications in peer-reviewed journals. He presented his findings at the American Peptide Symposium and was awarded the Young Investigator's Award. He is the author of *Application of Peptide-Based Prodrug Chemistry in Drug Development*, with a foreword written by Professor Jean Martinez (Former President, European Peptide Society) and published in the Springer series *SpringerBriefs in Pharmaceutical Science & Drug Development.* His research interests lie at the intersection of chemistry and medicine. Besides biomedical research, he is also generally interested in public health policy and general scientific outreach.

Chapter 1
Immune Homeostasis: Activation and Downregulation of NF-κB

1.1 Introduction to Immune System

Food is the ultimate source of consumable energy to sustain life on this planet. Every organism is a source of nutritious organic molecules, and could be potentially used up as "food" by their counterparts. Thus, it is conceivable that any organism that could defend against others would have a considerable selective advantage since it would avoid being "eaten up" by others (predators and/or pathogenic microorganisms). Thus, over the course of time, we have evolved various sophisticated "immune-defense mechanisms" to protect ourselves from other organisms, both large and small.

While coevolution of multicellular organisms and microbes have provided several essential mutual benefits for both [1], infectious diseases constitute a major threat to human health and a possible cause of death. Besides, it is also a major burden for the global economy. The sources of these infections are pathogenic organisms that include bacteria, viruses, or other parasites. Hence, it is very important for the host to be able to respond to such harmful threats, and indeed our ability to clear these microbial infections is vital for survival. However, this is not easy given the enormous diversity of pathogens, and their ability to mutate, evolve, and adapt rapidly to evade possible detection by the immune system. Hence, multicellular organisms have evolved several sophisticated immune recognition mechanisms to "fight" pathogens. Immune cells are present throughout the body. They are either found near the skin or gut where they can monitor the entry of foreign substances or discretely encapsulated in the spleen or thymus [2]. In vertebrates, the immune mechanism may be broadly classified as "innate" and "adaptive" [3].

A. De, *Ubiquitin Chains: Degradation and Beyond*, Springer Theses,
DOI 10.1007/978-3-319-14965-3_1

1.1.1 Innate Immunity

As suggested by the name, the innate immune system is evolutionarily ancient, and consists of anatomical barriers, immune cells, and protective proteins that are always present right from birth. The phylogenetically conserved innate immune system is the first line of defence, initially presenting a physical obstacle to microbes, and subsequently fighting the microbes at the site of infection. The main components of the innate immune system include physical barriers (skin and epithelial layers), phagocytic cells that "eat" microbes (neutrophils and macrophages), dendritic cells (critically important for activation of the "adaptive" arm of the immune system), natural killer (NK) cells, and circulating plasma proteins.

In order to recognize various molecules of microbial origin, the cells of the innate immune system (macrophage, dendritic cells) have germline-encoded receptors. The innate defense system is activated when these germline-encoded receptors recognize the molecular patterns that are conserved in a wide range of pathogens (PAMPs).

The first component of our immune system is the skin, which maintains a constant barrier against external infection. There are also other barrier tissues preventing harmful pathogens from getting inside the body. If the pathogens succeed in breaching the "first line of defense," and get into the body, the other cells of the innate immune system (macrophages, neutrophils) respond to the pathogens immediately, engulfing the foreign organisms by a process known as phagocytosis and killing them.

Thus, this innate, germline-encoded, one-size-fits-all immune system is always available and ready to fight any infection. In most cases, the actions of the innate immune system are sufficient to clear the infection. However, should the nonspecific innate response be insufficient to kill the pathogens, it is able to control the infection for a few days while simultaneously triggering the more specific adaptive immune response.

1.1.2 Adaptive Immunity

This is the more evolutionary recent immune arm first seen in jawed vertebrates. In general, the effector class of the adaptive immune response could be humoral (mediated by antibodies produced by B lymphocytes) or cell-mediated (mediated by T lymphocytes). A highly diverse group of T-cell and B-cell receptors generated by the process of VDJ recombination and somatic hypermutation form the basis of adaptive immune recognition. Receptors with appropriate specificities are clonally selected and expanded. "Clonal selection" forms the basis of "immunological memory," a process that lets the host "adapt" and remember the pathogen, enabling a more robust immunological response in case of a subsequent attack by the same pathogen. This is a major advantage for advanced organisms over their microbial

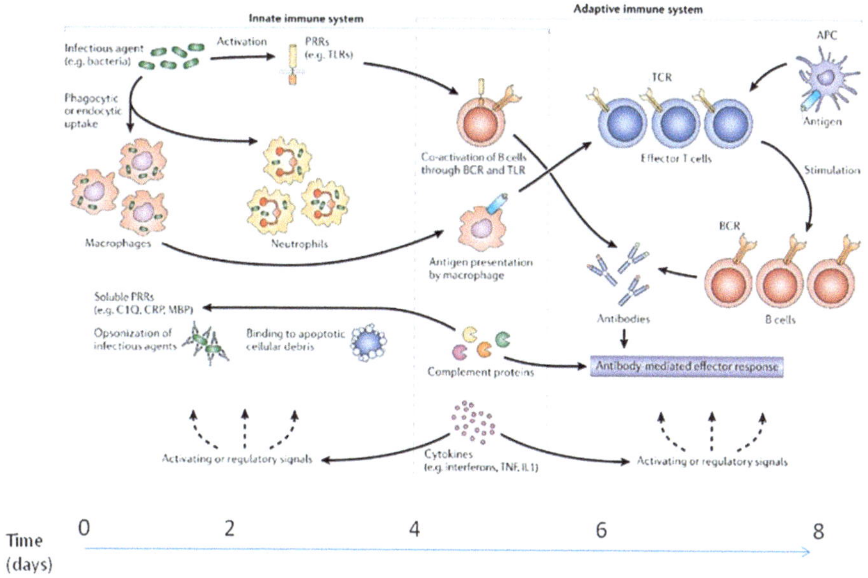

Fig. 1.1 Innate and adaptive immune system (*APC* antigen-presenting cell, *BCR* B-cell receptor) (Reprinted by permission from Macmillan Publishers Ltd: [Nature Reviews Genetics] (Gregersen PK, Behrens TW (2006) Genetics of autoimmune diseases--disorders of immune homeostasis. *Nat Rev Genet* 7: 917-928) copyright (2006))

counterparts. However, there are two limitations of an adaptive response. Firstly, "clonal selection" or the expansion and differentiation of specific lymphocyte clones to effector cells takes up to 4–7 days. Hence, no adaptive response can be mounted till this time, a considerable span for multiplication of fast replicating microorganisms. Secondly, since these receptors are formed randomly, they cannot by itself distinguish pathogenic antigens. Hence, the adaptive immune system does not function independently and innate immune signals activate and control the extent of the adaptive response.

Figure 1.1 shows the mechanisms of action of the innate and adaptive immune system. Innate immunity involves immediate, nonspecific response to pathogens. Thus, pattern-recognition receptors (PRRs) on macrophages and neutrophils recognize pathogen-associated molecular patterns (PAMPs) in pathogens leading to phagocytosis and opsonization. This leads to the secretion of cytokines like tumor necrosis factor α (TNFα), and interleukin-1β (IL-1β) that mediate the inflammatory response. These initial defensive acts by the innate immune system trigger the adaptive immune system. The lymphocytes (T cells and B cells) of the adaptive immune system have receptors that bind to specific antigens. The full development of the adaptive response requires selection, expansion, and differentiation of the specific responder cells. This takes time (4–7 days), but leads to memory lymphocytes which remember the specific antigen they responded to [4].

Although Elie Metchnikoff first described the innate immune system a century ago, the research in this field has been largely eclipsed by the intriguing discoveries of the many facets of the adaptive immune system. During this time, it was thought that the innate immune system only helps to keep the infection at bay while the adaptive immune system prepares to mount the specific and robust response. However discoveries in the last 20 years have clearly shown that the innate immune arm not only plays an essential role in detecting the infection and providing the immediate response, but also in simultaneously coordinating the initiation and determining the effector class of the adaptive response [5].

The fundamental strategy by which the innate immune system recognizes pathogens is by detecting those unique constitutive and conserved microbial molecules that are absent in the host. These molecules are typically part of essential microbial metabolic pathways and are required to sustain microbial life. Thus, it is impossible for microbes to live without expressing these gene products. By evolving to recognize these critical microbial products, the host gains a significant advantage. Some of these substances include lipopolysaccharide (LPS), peptidoglycan, and lipoteichoic acids (LTAs); all molecules found only in bacteria and not produced by eukaryotic cells. Hence, the hosts see these molecules as "signatures of the pathogen" and the recognition of such foreign microbial signatures is the first signal of an infection. While different strains and species of a common class of microbes (bacteria) may have minor variations in chemical structure (O-antigen subgroup of LPS is different in different bacterial species); the common invariant "molecular pattern" is what is recognized by the host innate immune system (main lipid-A pattern of LPS). Since the targets of innate recognition is invariant and conserved, they are known as PAMPs. The receptors of the innate cells that are responsible for recognition of PAMPs are known as PRRs, a broad class of receptors that have evolved over time and this initial recognition is primarily responsible for the specificity in immune response against the non-self [6, 7].

In order to recognize these different PAMPs, a number of PRRs have evolved in order to broadly sense the diversity of the microbial world (Fig. 1.2). The PRRs directly engage the evolutionarily conserved PAMPs to constitute what is the first step of the innate immune response against invading microbes. The different PRRs may be expressed on the surface of cells, in intracellular compartments, or secreted into blood or tissue fluids. PRRs may be broadly classified into, (a) transmembrane PRRs and (b) cytosolic PRR.

Transmembrane bound-PRRs include the Toll-like receptors (TLR) family which are probably the most well known as well as the best characterized of the PAMPs. The C-type lectin receptors (CLR) are another example of transmembrane receptors. The mannose receptors is a typical CLR expressed in macrophages which binds certain sugar molecules of some invading bacteria and viruses (this is especially important in the response against human immunodeficiency virus or HIV) [8].

Cytosolic PRRs recognize intracellular PAMPs. This is especially important when the invading microorganism has gotten inside the cell, either by itself or as a result of phagocytosis. They include CATERPILLAR/NOD-like receptors (NLRs) and RIG-l-like receptors (RLRs). The NLR family has around 20 members and bind

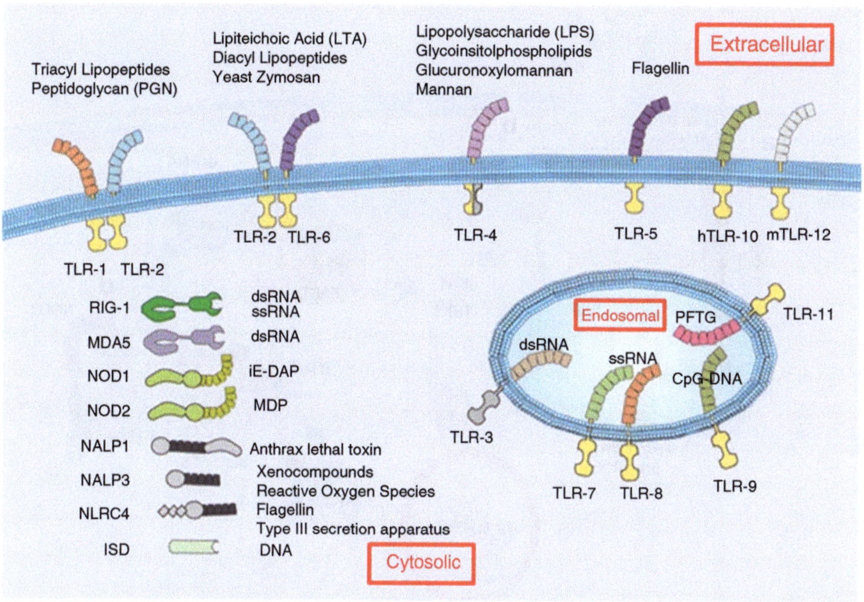

Fig. 1.2 Pattern-recognition receptors and their cognate ligands: PRRs may be membrane bound or endosomal in order to recognize various PAMPs in different locations (Reprinted by permission from Macmillan Publishers Ltd: [Cell Research] (Hayden MS, Ghosh S (2011) NF-kappaB in immunobiology. *Cell Res* 21: 223-244) copyright (2011))

to the peptidoglycans present in bacterial cell wall [9–11]. The *RLRs* play a crucial role in the immune response against viruses. Three different RLRs have been discovered, namely, RIG-I, MDA5, and LGP2. They sense viral replication in the host cytoplasm by interaction with dsRNA of viruses (host RNA is single stranded; a clear example of how the host immune system has evolved to distinguish between subtle differences in structures of "self" and "non-self") [12].

As we see, our body has an elaborate mechanism to respond to the constant challenge of various pathogenic microorganisms which trigger the immune system. The essential first line of defense involves the sentinel macrophages, sensing the various PAMPs with their PRRs. Remarkably, while binding of different PAMPs to their specific PRRs may trigger different intracellular signaling cascades, they all converge and lead to the activation of the nuclear factor κB (*NF-κB*) pathway (initially inactive in resting conditions) and the consequent expression of many proinflammatory cytokines [13] (Fig. 1.3). These proinflammatory cytokines are largely responsible for initiating the innate immune response by recruitment of additional professional phagocytes and subsequent pathogen clearance. They also shape the subsequent adaptive response [14]. One consequence of multiple pathways coalescing on a single transcription factor would be that dysregulation of NF-κB could be very harmful. Indeed, there are many reports associating dysregulation of NF-κB with different pathologies [15–17]. To avoid autoimmunity, it is essential that the

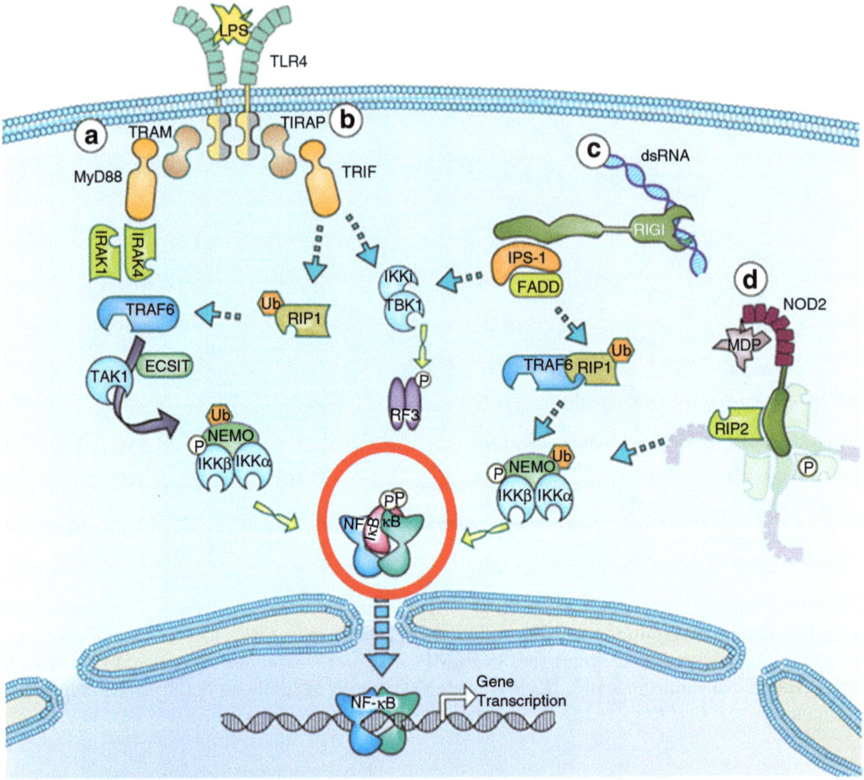

Fig. 1.3 Binding of PRRs to PAMPs converges on the NF-κB pathway via the activation of IKK complex (thus, signaling through the (*a*) TLRs (Myd88) (*b*) TLRs (TRIF) (*c*) RIGI (*d*) NOD pathogen-associated molecular patterns lead to NF-κB activation) (Reprinted by permission from Macmillan Publishers Ltd: [Oncogene] (Hayden MS, West AP, Ghosh S, (2006) NF-kappaB and the immune response. *Oncogene* 25: 6758-6780) copyright (2006))

activation of NF-κB be tightly regulated in order to limit the duration and magnitude of the response. This thesis deals with an important inhibitor of NF-κB: A20 (also known as TNFAIP3) [18].

1.2 NF-κB: An Overview

As seen in the last section, the mammalian immune response can be broadly subdivided into innate and adaptive responses. The first response is provided by the innate system and begins with host-recognition of the pathogen. The subsequent responses take place at various levels of complexity at the cellular, tissue, and organismal levels. The ultimate aim is to clear the pathogen. From a basic molecular point of view, it is insightful to distill the immune response into individual signal transduction events which alters gene expression at the cellular level, finally leading to a concerted immune response at the organismal level [19]. NF-κB, inactive in the basal state and activated upon sensing PAMPs, is one key transcription factor that plays a key role in mediating transcriptional changes. The gene products of the initial response include cytokines acting as "messengers" of the "initial recognition of pathogens," thereby propagating the immune response. Additionally, these cytokines can also activate NF-κB, thereby focusing and elaborating the immune response. Thus, we find that NF-κB mediates critical aspects of the innate and adaptive responses in a surprisingly large number of cases. The importance of this inducible transcription factor can be gauged from the volume of research done in the past 25 years since it was identified [20] as a nuclear factor interacting with immunoglobulin enhancer sequences [21, 22]. A search in PubMed using "NF-κB" alone results in more than 40,000 hits.

While the underlying interest in studying NF-κB comes from its importance in health and disease, these studies also serve as a model for studying a wide range of biological responses that depend on inducible transcription factors. Inducible gene expression plays an important role in both prokaryotes and eukaryotes. It is an important regulator of normal physiology as well as the key in allowing multicellular organisms to adapt to chemical, environmental, and pathogenic stresses. Fundamental biological processes including organ morphogenesis and differentiation of both single cells and multicellular organisms depend on the paradigm of inducible gene expression.

NF-κB has been found to be amenable to a wide range of experimentation at the biochemical, cellular, and organismal levels. Most of the studies of NF-κB have been done from an immunological background, as the transcription factor plays an essential and evolutionarily conserved role in responding to immune insults. However, while much of our understanding of the pathway comes from studies in immunology, it is important to appreciate that NF-κB plays an important broader role in regulating gene expression that affects cell survival and apoptosis, differentiation, and proliferation [23, 24].

A brief overview of the NF-κB pathway may be helpful here. NF-κB is bound to inhibitory proteins (IκB) and is inactive in the unstimulated state. The pathway is activated by inducing stimuli acting through receptors and adapter proteins to trigger IκB-kinase complex (IKK) (activating kinase) activation. This leads to the phosphorylation, ubiquitination, and subsequent degradation of IκB. NF-κB is now free and undergoes a series of posttranslational modifications to be fully activated (Fig. 1.4). It translocates to the nucleus, binds to specific DNA sequences, and turns on the transcription of its target genes. Thus, the basic components of the pathway are the receptors and adaptor molecules, IKK complex, IκB proteins, and NF-κB subunits. The pathway is controlled by multiple positive and negative regulatory elements [24].

In mammals, the NF-κB family of transcription factors is comprised of five constituent monomers: p50, p52, p65 (also RelA), c-Rel, and RelB (Fig. 1.5). All of them have a N-terminal DNA-binding domain which is also the domain necessary for homodimer/heterodimer formation. This N-terminal domain is also known as

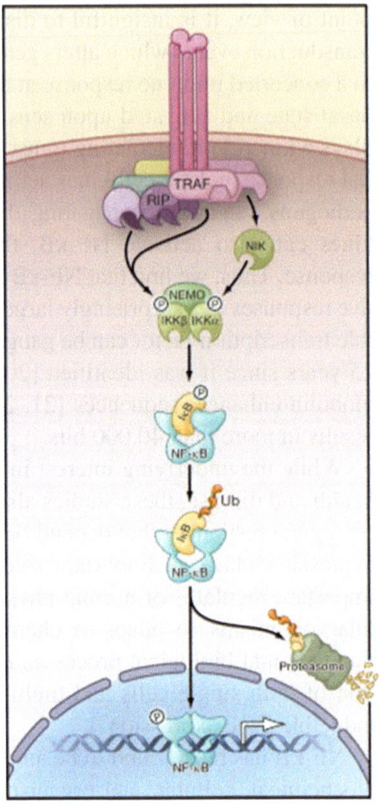

Fig. 1.4 Basic NF-κB signaling pathway (Reprinted from Cell, 132, Hayden MS, Ghosh S, Shared principles in NF-kappaB signaling, 344-362, Copyright (2008), with permission from Elsevier)

the Rel homology domain (RHD). The crystal structures of multiple NF-κB dimers reveal that the RHD is comprised of two immunoglobulin-like folds. Of them, one is engaged in dimerization while the other is responsible for recognition of specific sites on the DNA. Minor differences in dimer interfaces determine partner preferences and selectivity in DNA binding [25]. Multiple posttranslational modifications (phosphorylation and acetylation) of different residues in these molecules further modulate DNA binding [26]. Subsequent transcriptional activities may additionally depend on the interaction of dimers with other activator proteins [27]. Activated NF-κB dimers can bind multiple, related DNA sequences called κB promoter sites to turn on the expression of different genes. The C-terminal transcription activation domains (TADs) are necessary for activator recruitment and consequent transcription of target genes. Only RelB, c-Rel, and p65 contain the TAD. Since p50 and p52 lack the TAD, they can act as transcriptional activators only upon association with RelB, c-Rel, and p65. However, p50 and p52 can form homodimers that bind to the DNA, and repress transcription. In order to positively regulate transcription, the p52 and p50 subunits have to form heterodimers with the other factors that have TAD [24]. RelB also has a unique leucine zipper (LZ) motif in its N-terminus which plays an important role in regulating transcription.

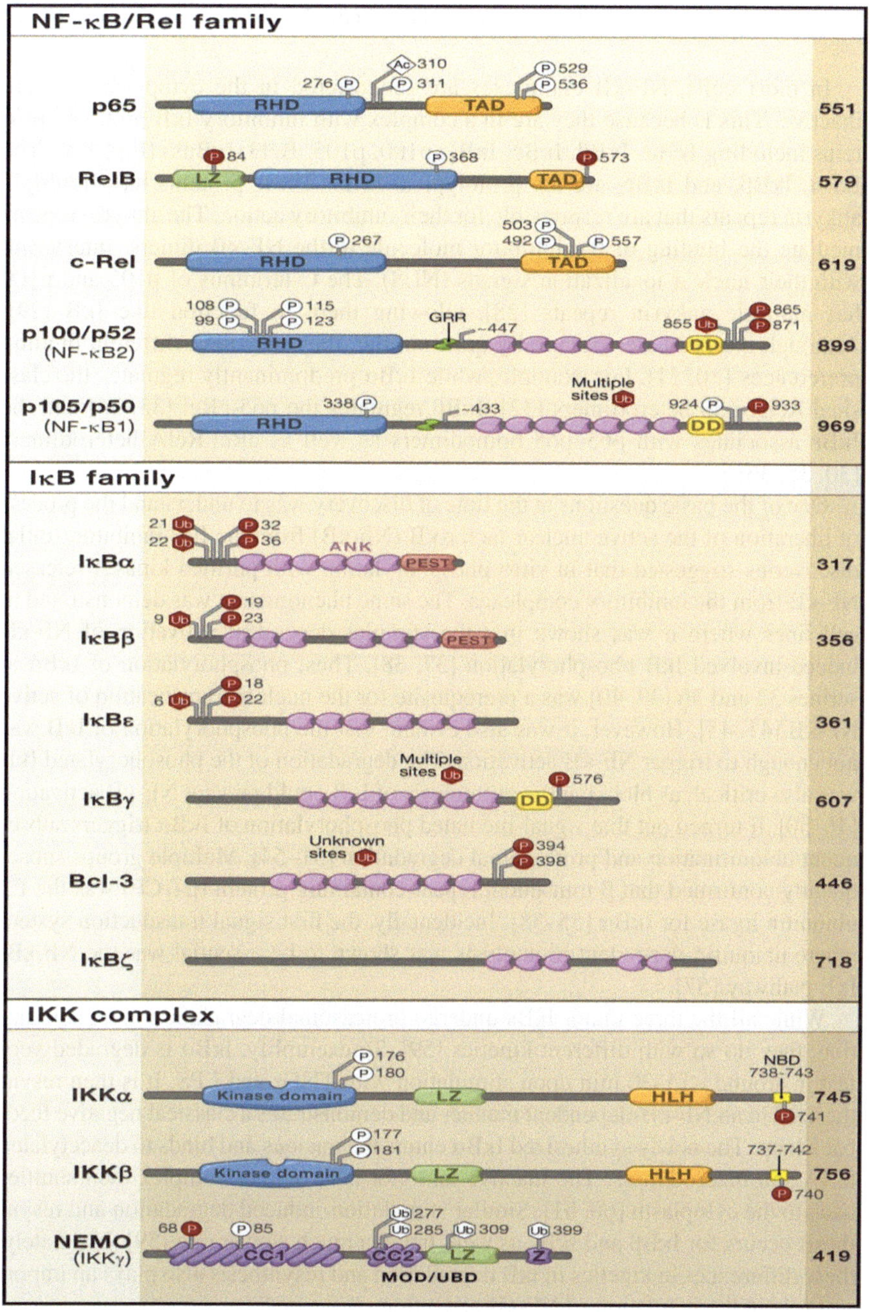

Fig. 1.5 Mammalian NF-κB, IκB, and IKK protein families: Phosphorylation, ubiquitination, or acetylation are indicated with P, U, or Ac respectively. Inhibitory events or degradative phosphorylation and ubiquitination sites on p100, p105, and IκB proteins are indicated by *red* Ps and Us, respectively (*RHD* Rel homology domain, *TAD* transactivation domain, *LZ* leucine zipper domain, *GRR* glycine-rich region, *HLH* helix-loop-helix domain, *Z* zinc-finger domain, *CC1/2* coiled-coil domains, *NBD* NEMO-binding domain, *MOD/UBD* minimal oligomerization domain, and ubiquitin-binding domain, *DD* death domain) (Reprinted from Cell, 132, Hayden MS, Ghosh S, Shared principles in NF-kappaB signaling, 344-362, Copyright (2008), with permission from Elsevier)

In most cells, NF-κB complexes are sequestered in the cytoplasm and are inactive. This is because they are in a complex with inhibitory IκB proteins (proteins including IκBα, IκBβ, IκBε, IκBζ, p100, p105, Bcl3, IκBns) (Fig. 1.5). The IκBα, IκBβ, and IκBε, are the prototypical IκBs. These proteins have multiple ankyrin repeats that are responsible for their inhibitory action. The ankyrin repeats mediate the binding of the inhibitor molecule to the NF-κB dimers, interfering with their nuclear localization signals (NLS). The C-terminus of p105 and p100 has multiple ankyrin repeats [28], allowing them to function like IκB [29]. Although the IκBs are structurally quite similar, they may have different binding preferences [30, 31]. For example, while IκBα predominantly regulates the classical RelA-p50 heterodimers [32], IκBβ regulates the p65-cRel [33, 34] and the IκBε associates with p65–p65 homodimers as well as cRel:RelA heterodimers [30, 35, 36].

One of the basic questions at the time of discovery was to understand the process of liberation of the active nuclear factor-κB (NF-κB) from the IκB inhibitor. Initial discoveries suggested that in vitro phosphorylation with purified kinases released NF-κB from the inhibitory complexes. The same phenomenon was demonstrated in cell lines where it was shown that the stimulus-dependent activation of NF-κB indeed involved IκB phosphorylation [37, 38]. Thus, phosphorylation of IκBα at serines 32 and 36 [39, 40] was a prerequisite for the nuclear translocation of active NF-κB [41–45]. However, it was also evident that the phosphorylation of IκB was not enough to trigger NF-κB activation. The degradation of the phosphorylated IκB was also critical, as blocking the proteolysis of IκB could prevent NF-κB activation [46–50]. It turned out that signal-mediated phosphorylation of IκBα triggers subsequent ubiquitination and proteasomal degradation [51–54]. Multiple groups subsequently confirmed that β transducin repeat-containing protein (βTrCP) was the E3 ubiquitin ligase for IκBα [55–58]. Incidentally, the first signal transduction system where ubiquitin-dependent proteolysis was shown to be essential was the NF-κB/IκB pathway [57].

While all the three major IκBs undergo proteasomal degradation upon stimulation, they do so with different kinetics [59]. To exemplify, IκBα is degraded very fast in around ~15–30 min upon stimulation with TNFα and LPS. It is then resynthesized in an NF-κB-dependent manner and demonstrates a classical negative feedback loop. The newly synthesized IκBα enters the nucleus and binds to deacetylated RelA:p50 heterodimers. The inactive inhibitor-heterodimer complex then shuttles back to the cytoplasm [60, 61]. Similar stimulation-induced degradation and resynthesis occurs for IκBβ and IκBε as well, but at a much slower rate [59]. Ultimately, these differences in kinetics in IκB degradation and resynthesis also plays an important role in the regulation of NF-κB activation.

The kinase responsible for phosphorylation of IκB in cells was later identified and named the IKK, which was shown to have both catalytic (IKKα and IKKβ) and regulatory (IKKγ/NEMO) subunits [62–66]. Based on the diversity of stimuli activating NF-κB and the number of IκBs, one could have expected that multiple kinases were responsible for phosphorylation of the different IκBs. As it turned out, only one kinase complex is responsible for the phosphorylation. Besides the traditional role in IκB phosphorylation, the IKK complex has also been shown to enhance the transcriptional activity of NF-κB by phosphorylating the TADs of cytoplasmic RelA and c-Rel while still in the cytoplasm [26, 67].

1.3 NF-κB: The Canonical Pathway and Non-canonical Pathway

There are two distinct pathways for NF-κB activation, the canonical pathway and the non-canonical pathway. They are activated by different ligands and kinases. The canonical pathway mediates the typical inflammatory responses, while the non-canonical pathway plays a role in the slow differentiation and maturation of immune cells and secondary lymphoid organogenesis. The canonical pathway is dependent on NEMO, while the non-canonical pathway does not depend on it [68] (Fig. 1.6).

While the two pathways are quite distinct, there are evidences of crosstalk between them [68]. The pathways are briefly described here:

(1) *Canonical pathway*: The canonical *NF-κB* signaling pathway is activated by proinflammatory ligands that include PAMPs, cellular danger-associated molecular patterns (DAMPs), and cytokines. The effector NF-κB heterodimers activated downstream are mainly the p65:p50 and c-Rel:p50 heterodimers. Upon ligand binding to cognate receptors, various adapter proteins trigger the activation of IKK complexes containing the regulatory NEMO subunit. Deletion of NEMO causes massive liver degeneration resulting in embryonic lethality [69], demonstrating the importance of this pathway and NEMO for life. While this pathway is characterized by the essential requirement of the regulatory NEMO subunit, the catalytic subunit of IKK that is activated and responsible for phosphorylating the inhibitory IκB is IKKβ. IKKβ is activated by phosphorylation of its serine residues at positions 177 and 181 [70]. Additional evidence for the signaling axis of NEMO-IKKβ-p65 comes from the uncanny similarity of the phenotypes of the IKKβ knockout [71] and the RelA knockout [72], both of which resemble the NEMO knockout and show apoptosis and degeneration in the liver.

A brief description of the pathway is provided here. The p65:p50 dimer is sequestered in the cytoplasm by the binding of IκBα. The IκBα:RelA:p50 trimer is mostly localized to the cytoplasm as the IκBα has a strong nuclear-export signal. However, this complex also shuttles between the nucleus and cytoplasm [73, 74]. This is because the binding of IκBα to p65:p50 masks the NLS of p65 but fails to mask the NLS of p50 [32].

Upon stimulation, the activated IKK phosphorylates IκBα on Ser32 and Ser36 [75]. The phosphorylated IκBα is now polyubiquitinated by degradative Lys48-linked polyubiquitin chains at Lys19 by the Skp1, Cdc53/Cullin1, and F-box protein βTRCP $SCF^{I\kappa B}$ E3 ubiquitin ligase complex [32]. This leads to the degradation of ubiquitinated IκBα via the 26S proteasome. The strong p65 NLS is now exposed and the p65:p50 dimers translocate to the nucleus. In addition to IκBα, IκBβ and IκBε are also substrates for IKK phosphorylation and subsequent degradation via

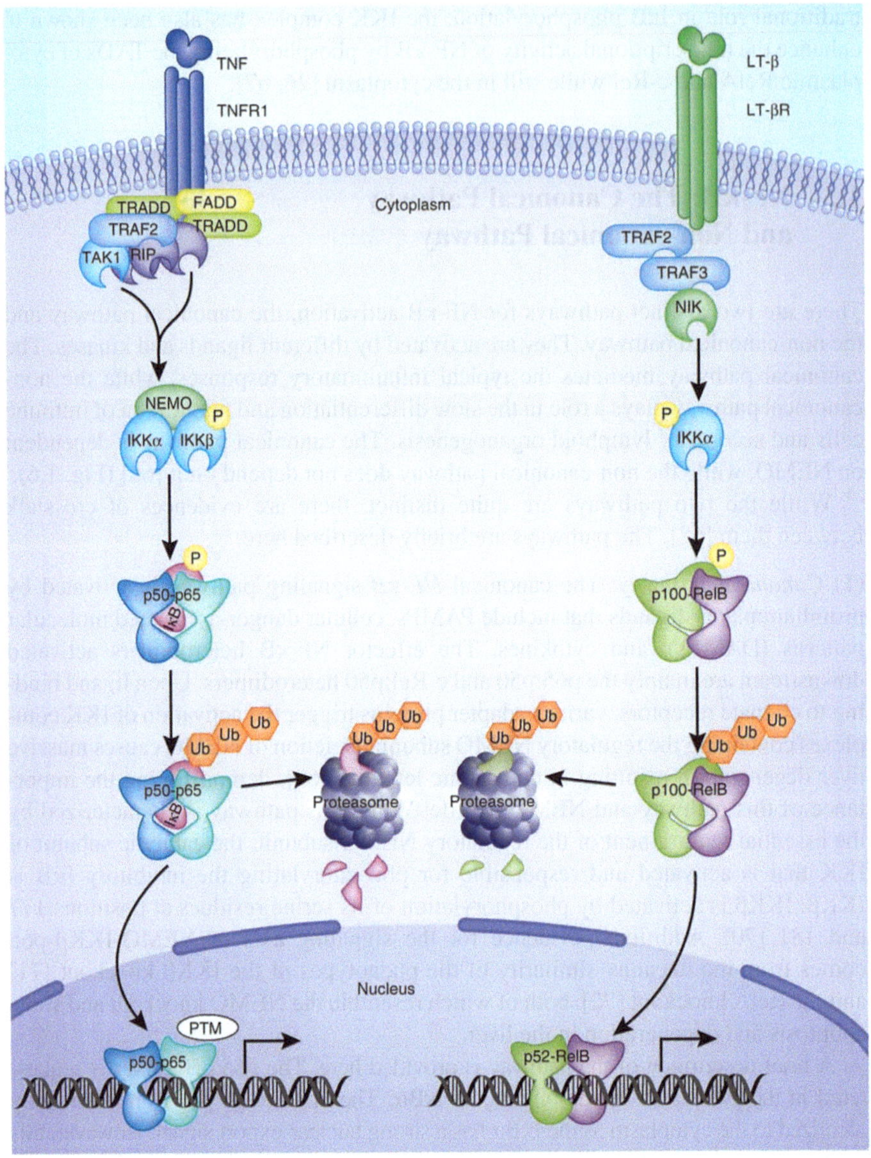

Fig. 1.6 Canonical and non-canonical NF-κB signaling (Reprinted by permission from Macmillan Publishers Ltd: [Nature Immunology] (Oeckinghaus A, Hayden MS, Ghosh S (2011) Crosstalk in NF-kappaB signaling pathways. *Nat Immunol* 12: 695-708), copyright (2011))

similar pathways [76]. That being said, the affinity of IKK to the various IκBs is markedly different. Hence, the various IκB protected dimers are degraded with different kinetics. The net consequence is that the basally inactive NF-κB transcription factors are inducibly activated by a diverse set of stimuli, all leading to the nuclear translocation and subsequent expression of NF-κB dependent genes [77, 78].

(2) *Non-canonical pathway*: This alternative NF-κB pathway is activated upon stimulation by a small subset of TNF family members that are involved in differentiation, maturation, and development of immune cells. These include lymphotoxin β, LTαβ, BAFF (B cell-activating factor), CD40 ligand, RANKL (receptor activator of NF-κB ligand), and TWEAK (TNF-related weak inducer of apoptosis) [79–82]. Since developmental processes require sustained signaling, it is not surprising that the kinetics of activation and deactivation of this pathway is slower as compared to the canonical pathway. Hence the pathway has a long-lasting activity, as opposed to the rapid transient activation of p65–p50 heterodimers by inflammatory stimuli.

Biochemical characterization has shown that in unstimulated cells, RelB interacts with and is associated with p100. The RelB:p100 complex is stabilized by multi-domain interactions. Thus, p100-deficient cells show reduced levels of RelB and similarly, RelB knockout cells have less p100 protein [83, 84]. This alternative NF-κB pathway ultimately leads to the inducible proteasomal processing (as opposed to degradation) of p100 to p52 and activation of RelB:p52 heterodimers. Another major distinction with the canonical pathway lies in that this pathway does not require NEMO. Instead, the non-canonical *NF-κB* pathway is activated strictly through a NEMO-independent IKKα-mediated pathway [78, 81, 85, 86]. Ligand binding activates the NIK protein, which directly phosphorylates p100 at Ser866 and Ser870, and also phosphorylates IKKα [86, 87]. The activated IKKα then phosphorylates p100 in its ankyrin domain at Ser99, Ser108, Ser115, Ser123, and Ser872 [76, 88]. The SCG/βTRCP E3 ligase complex then recognizes the phosphorylated p100 and polyubiquitinates it with K48-linked ubiquitin chains [88, 89]. This leads to the consequent proteasomal processing of p100 by the 26S proteosome [86, 87, 90] to p52. The p52 then associates with the RelB that was previously associated with p100 to form the activated RelB:p52 dimers.

Elegant genetic proof of NIK, IKKα, and RelB as components of the non-canonical pathway comes from the remarkable similarity of the phenotypes of mice deficient for these genes [85, 91, 92]. They all show defective development of Peyer's patches and other lymph nodes. Additionally, the phenotype of *aly/aly* mice (carries a point mutation in kinase domain of *Nik* gene leading to inactivation of NIK) is similar to *IkkαAA* mice (mice have catalytic serines mutated to alanine; as a result IKKα cannot be activated in these mice) [86, 93].

1.4 A Brief Overview of the IKK Complex

The NF-κB pathway is activated by a whole range of extracellular ligands (both membrane bound and soluble). They include members of the TLR, TNFR, IL-1R superfamilies. Additionally, NF-κB has been recently shown to respond to changes in the intracellular environment as well. These changes may be a consequence of DNA damage, elevated reactive oxygen species, and recognition of pathogens by the RIG-I and NOD family of proteins. All of these stimuli converge on the activation of the IκB kinase (IKK) complex.

The IKK complex comprises of two catalytic subunits (IKKα and IKKβ) and one regulatory subunit (NEMO or IKKγ) (Fig. 1.5). There is a 52 % overall sequence identity between the catalytically active kinase subunits IKKα and IKKβ, while the catalytic domain are 65 % identical. Amongst the two catalytic subunits of IKK, IKKβ contributes more than IKKα towards IκB kinase activity in most cell types and thus seems to be the more important catalytic subunit. Loss of IKKβ leads to lethality, thus IKKα cannot compensate for the IκB kinase activity in absence of IKKβ [94] However, loss of IKKα has little effect on net IKK activity as IKKβ can compensate for IKKα [95].

There is a plethora of clear genetic evidence pointing to the essential role of all the three subunits of IKK [96]. Mice lacking IKKβ resemble p65 knockouts. This is expected since IKKβ plays a profound role in the canonical activation of p65-containing dimers [71, 94, 97]. The embryonic lethality can be rescued by deletion of TNFR1, demonstrating a crucial role for IKKβ in TNF-mediated signaling [71, 94, 98]. IKKα knockout mice die soon after birth because of severe morphological defects in the skin and the limbs [95, 99, 100]. Initially, there was little evidence for IKKα playing a role in NF-κB activation. However, subsequent reports demonstrated the central importance of IKKα in the non-canonical NF-κB pathway and in certain canonical pathways as well [101]. NEMO is essential in the canonical pathway; hence NF-κB cannot be activated via the canonical pathway in NEMO-deficient cells. In addition, mice lacking NEMO die from severe liver degeneration [66, 69, 102].

It is now understood that almost all NF-κB-stimulating ligands act via the IKK complex [103] as there is no NF-κB activity mice lacking both IKKα and IKKβ [103]. Surprisingly, many of the components upstream of IKK are similar, and also show significant mechanistic overlap (even in pathways that are functionally divergent). However, even 15 years after the initial characterization of the IKK complex, it is still not entirely known how IKK gets activated. One of the major reasons for this is that, in general, the receptors that lead to IKK activation lack enzymatic activity [104].

What is known is that the activation of IKK depends on the phosphorylation of serines in the activation loop of the IKKs. Thus IKKα is activated upon phosphorylation of Ser 176 and Ser 180 while IKKβ is activated upon phosphorylation of Ser 177 and Ser 181. This phosphorylation of active loop serines is essential for inducing kinase activity of the IKK complex. Treatment with phosphatases decreases kinase activity in vitro, and mutating the serines to alanines abolishes signal responsiveness. Conversely, mutation to glutamic acid results in IKK that is constitutively active [63, 70, 105–107].

Thus, although receptors leading to IKK activation lack kinase activity, it is evident that some kinase has to be responsible for phosphorylating and activating IKK. As described before, there is both biochemical and genetic evidence for NIK acting as the kinase to directly phosphorylate and activate IKKα in the non-canonical pathway [93, 107, 108]. For the canonical pathway, it is hypothesized that IKK could either be activated by autophosphorylation or phosphorylation by an upstream kinase [104]. While multiple kinases can act as an IKK kinase in vitro (IKK-K), they have all failed the genetic test. It has been found that many molecules that have kinase activity act as adaptor proteins in the activation of IKK. Currently, the only acceptable candidate kinase to phosphorylate IKK is TAK1. Indeed, the failure of multiple potential kinases to satisfy the gene-knockout test has supported the theory that IKK might activate itself by autophosphorylation.

Recent resolution of the crystal structure of IKKβ suggests that a dimer of IKKβ could not phosphorylate itself. This is because the active site of the one IKKβ would be distant from the "serine activation loop" of the second IKKβ of the dimer [109]. However, the active site and activation loop are much closer in related higher-order structures that would mediate IKK activation [109]. Hence, while it seems that IKKβ could not undergo *cis*-autophosphorylation; there is indeed a possibility that two dimers could autophosphorylate and activate each other by a process known as *trans*-autophosphorylation [110]. For this to happen, the two dimers need to be in close induced proximity with the right conformation [70, 111].

These recent structural insights are in line with documented evidence suggesting that large oligomeric signaling complexes need to assemble for IKK activation. Thus, NF-κB activation might be mediated by recruitment of IKK to the receptor by adaptor proteins and signaling complexes from the TRAF/RIP (receptor-interacting protein) family members [70, 111–113]. Upon recruitment of the IKK complex to the receptor, it either activates itself by autophosphorylation or is phosphorylated by another kinase like TAK1.

Different pathways leading to IKK activation share multiple signaling intermediates upstream of the IKK complex. These molecules are often the non-kinase adaptor molecules like the RIP and TRAF proteins. It is still an open question whether IKK activation in the canonical pathway is mediated by autophosphorylation or TAK-1. However, there is no doubt that a kinase-independent organization of adaptor molecules leading to formation of receptor signaling complexes is necessary for either of these possibilities and therefore, in the downstream activation of NF-κB [104].

1.5 Ubiquitination and NF-κB

Ubiquitin is a small protein made of 76-amino-acid that is covalently linked to its target protein [114, 115]. The protein is highly conserved across eukaryotes from yeast to humans. Ubiquitin is encoded in the genome as "ubiquitin precursors" with ubiquitin fused with a ribosomal subunit or two ubiquitins fused together. There are four such genes encoding for "ubiquitin precursors" [116]. Ubiquitination may affect a protein's half-life, localization, or function. Ubiquitination is a post-translational modification by which the epsilon amine of a particular lysine of a protein can be covalently linked through an isopeptide bond with the carboxylic acid of the C-terminal glycine of ubiquitin. The process occurs in three consecutive steps catalyzed by three classes of enzymes. These enzymes are the Ub-activating enzymes (or E1), Ub-conjugating enzymes (or E2), and Ub ligases (or E3) [117] (Fig. 1.7).

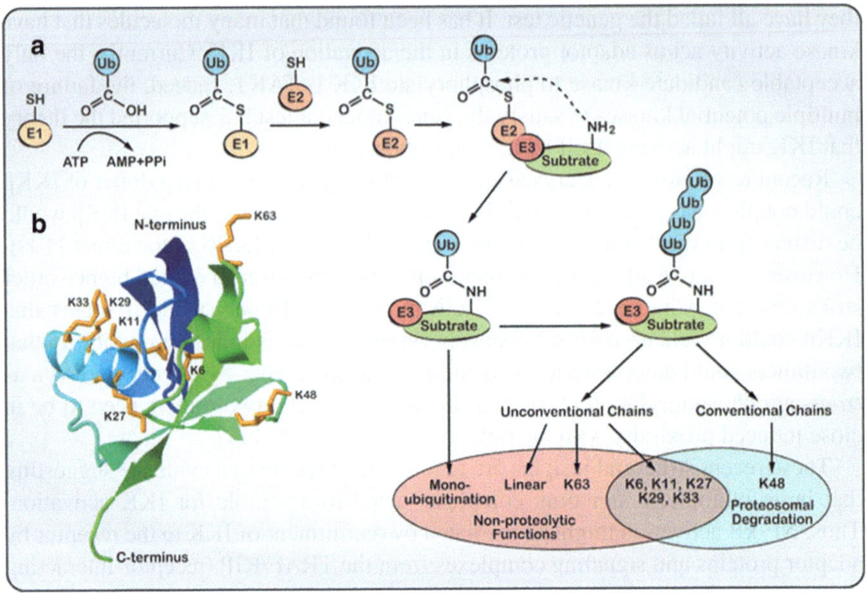

Fig. 1.7 Ubiquitination and functions. (**a**) Three-step cascade that leads to ubiquitination; functions of diverse ubiquitin chains (**b**) Structure of ubiquitin (PDB code: 1UBQ), seven lysine residues are highlighted (Reprinted by permission from Macmillan Publishers Ltd: [Cell Research] (Liu S, Chen ZJ (2011) Expanding role of ubiquitination in NF-[kappa]B signaling. *Cell Res* 21: 6-21), copyright (2011))

There are only two E1 enzymes in the genome. They catalyze the transfer of a thio-ubiquitin intermediate to a cysteine residue of a specific E2 enzyme. This transfer is an energy-intensive process and proceeds by hydrolysis of an ATP. Thus ubiquitination requires energy and the cell uses it to mostly degrade its substrates. There are over 50 E2 enzymes which transfer the ubiquitin to an E3 ligase. There are around 700 E3 ligases and the specificity of ubiquitination partly comes from the E2–E3 interaction. Additionally, the E3 binds to the specific substrate and ligates the carboxyl terminus of the ubiquitin to the ϵ-amino group of the lysine (Lys) residue of that specific substrate [118]. Subsequently, multiple rounds of ubiquitination result in the formation of polyubiquitin chains. There are seven lysine residues in ubiquitin (K6, K11, K27, K29, K33, K48, K63), and any one of these can participate in elongation of the polyubiquitin chain. Thus, polyubiquitin chains may have different linkages. A linear ubiquitin chain may also be formed by attachment of N-terminal amine of one ubiquitin to the C-terminal carboxyl group of another ubiquitin.

The E2–E3 combinations play a major role in determining the specific linkage of the ubiquitin chains. For example, a particular E2 (UBC5) complexes with another E3 ligase ($SCF^{\beta TrCP}$) to trigger the formation of K48-linked polyubiquitin chains. Similarly, complexing of UBC13 (E2) with TRAF proteins (E3) forms a K63-linked polyubiquitin chain [119].

These different kinds of polyubiquitin linkages have different structures and topologies. Ubiquitin chains are recognized by proteins having an ubiquitin-binding domain (UBD). Thus, different proteins with distinct UBDs bind the specific ubiquitin linkages. There are more than 20 different types of UBDs that have been identified till now [115, 120, 121], differing substantially in their sizes (20–150 amino acids). For example, the proteasome subunits Rpn13/ARM1 and Rpn10/S5a have ubiquitin-associated (UBA) domains that preferentially bind K48-linked chains, while NEMO has a UBAN (UBD in ABIN and NEMO) motif that binds with both K63 and linear polyubiquitin chains (but not K48) [122]. Hence, different ubiquitin chains can potentially confer different fates to the substrate protein (Fig. 1.7). K48-linked ubiquitination (and K11) is known to dictate proteasomal degradation, while K63-linked chains have been proposed to function as scaffolds to assemble the different signaling complexes, thereby regulating different processes like DNA repair, chromatin remodeling, or activation of NF-κB. Thus, K63 ubiquitination is also known as regulatory ubiquitination [122–124]. UBDs can act as receptors for ubiquitin, thus in principle, it is possible for proteins that have UBDs to potentially transduce signals from substrates that are ubiquitinated to other components of the signaling pathway [115].

Like phosphorylation, ubiquitination is also a reversible posttranslational modification. Thus, ubiquitination can be reversed by deubiquitination. The process of deubiquitination is carried out by proteases specifically known as deubiquitinating enzymes or deubiquitinases (DUBs). There are close to 100 DUBs in the human genome [125]. These DUBS can be subdivided into five broad families by the presence of various protease domains: they may be metalloproteases, papain-like cysteine proteases (i.e., a carboxy-terminal hydrolase), ubiquitin-specific protease, Joseph disease proteases, or an ovarian tumor containing protease (OTU deubiquitinase).

The specificity of DUBs is determined by their UBDs and other protein interaction motifs. This directs them to specific ubiquitinated substrates containing a particular type of poly-Ub chain linkage [126]. If K63-linked polyubiquitin chains positively regulate NF-κB signaling, then their deubiquitinases would be expected to be critical regulators of the pathway. Indeed, A20 and CYLD have been reported to be two such critical proteins with deubiquitinase activity. However, what is not known is if the deubiquitinase activity of these proteins is actually responsible for their function, or if these proteins act by a different mechanism. Mice lacking A20 are perinatal lethal, suffering from uncontrolled multiorgan inflammation as a result of persistent NF-κB activity [127]. Additionally, mutations or organ-specific deletions of TNFAIP3 (the gene encoding for A20) are associated with different kinds of lymphomas [128–131]. Moreover, patients with the autoimmune disorder systemic lupus erythematosus also have been found to have polymorphisms in their *TNFAIP3* locus [132]. CYLD is another deubiquitinase that acts as a tumor suppressor and negatively regulates NF-κB signaling. Patients with familial cylindromatosis have been found to have multiple somatic mutations in CYLD and deletion of CYLD may cause colonic, hepatocellular, and renal carcinomas, in addition to multiple myeloma [133].

1.6 Degradative Ubiquitination in Canonical and Non-canonical NF-κB Signaling

Aaron Ciechanover, Avram Hershko, and Irwin Rose elucidated the ubiquitin pathway and characterized the various enzymes (E1, E2, E3) and deubiquitination enzymes (DUBs) [134, 135]. For this, they were jointly awarded the Nobel Prize for Chemistry in 2004. Subsequently Alex Varshavsky and colleagues showed that ubiquitination targets proteins degradation in vivo [136, 137]. While there is room for abundant skepticism for the role of K63-linked polyubiquitination, there is no doubt that K48-linked ubiquitin chains play a prominent role in NF-κB signaling. Activation of NF-κB is defined by IκB degradation or by the proteasomal processing of NF-κB precursor proteins and this is mediated by degradative ubiquitination [138] (Fig. 1.8).

Canonical NF-κB signaling may be defined as those pathways which lead to "degradation of IκB." These pathways are mostly activated by proinflammatory stimuli including bacterial PAMPs like LPS and cytokines like interleukin (IL)-1β and TNFα. Stimulation leads to the activation of the IKK complex which phosphorylates the IκB proteins. This targets the phosphorylated IκBs for K48-mediated polyubiquitination by a K48-specific ubiquitin ligase complex that consists of Skp1, Cul1, Roc1, and βTrCP [139]. The ubiquitinated IκB undergoes degradation by the 26S proteasome. This allows NF-κB to translocate to the nucleus and begin transcription of target genes.

The non-canonical pathway is defined as the pathway that leads to the "proteasomal processing" of precursor-p100 to its mature p52 subunit [138]. This happens when some receptors (such as B-cell activating factor receptor and CD40) are stimulated in B lymphocytes. Upon stimulation, NIK activates IKKα, which in turn phosphorylates p100. p100 is now ubiquitinated by the βTrCP ubiquitin ligase complex. Under most circumstances, K48-polyubiquitination results in complete proteasomal degradation. However the ubiquitinated p100 is not completely degraded but the C-terminal ankyrin repeats are selectively degraded by the 26S proteasome. This leaves the N-terminal portion intact, and this is the mature p52 subunit. This is because a glycine-rich region in the N-terminus of p100 prevents it from degradation [140]. Additionally, the N-RHD domain is tightly folded and forms a stable dimer, making this region resistant to degradation [141].

1.7 Oligomerization in Activation of IKK

A cursory look at any of the three major NF-κB activation pathways (TNF, Toll/IL-1, or T cell receptor) shows that oligomerization is a shared theme in NF-κB signaling [24] (marked in red in Fig. 1.9) and recruits the IKK complex to the receptor. Dimerization of the recruited IKK has been shown to be necessary and sufficient for activation of IKK [62, 111, 142].

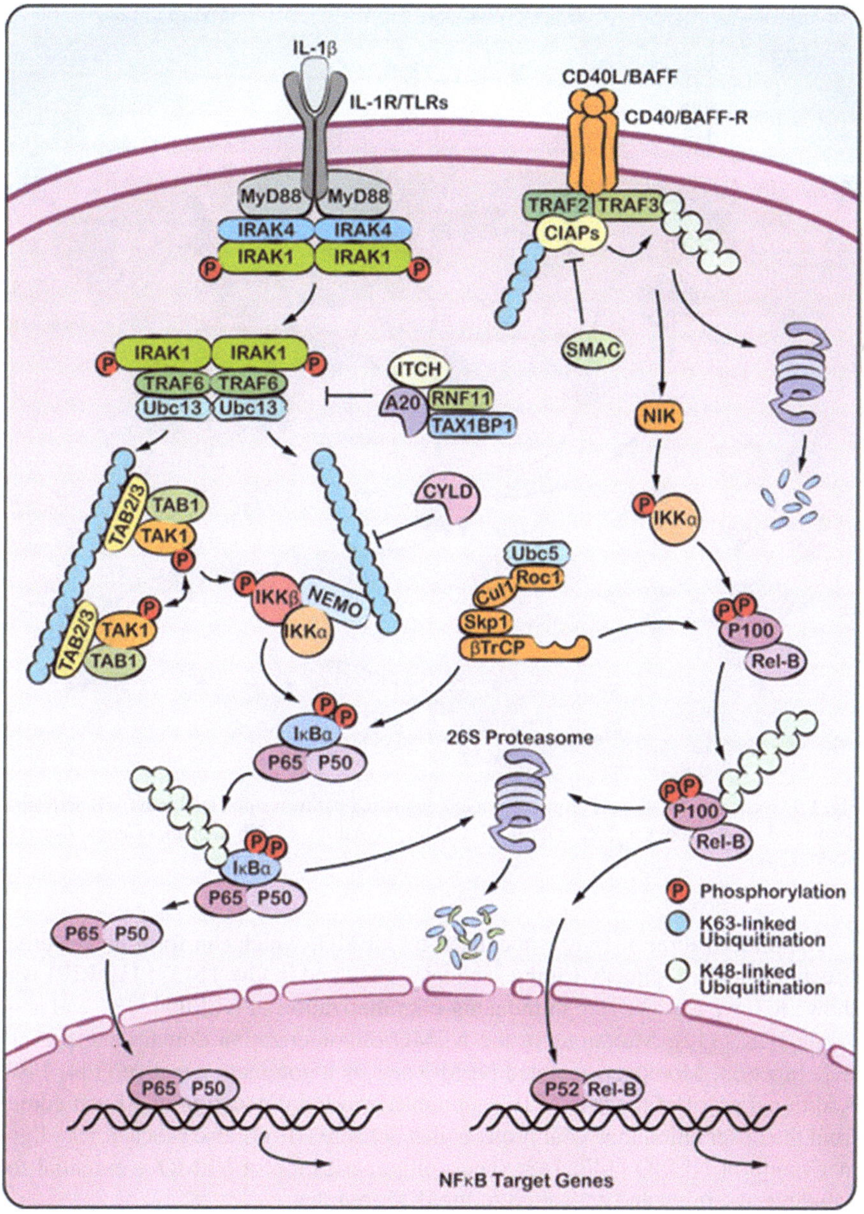

Fig. 1.8 Role of ubiquitination in the canonical and non-canonical NF-κB pathways (Reprinted by permission from Macmillan Publishers Ltd: [Cell Research] (Liu S, Chen ZJ (2011) Expanding role of ubiquitination in NF-[kappa]B signaling. *Cell Res* 21: 6-21), copyright (2011))

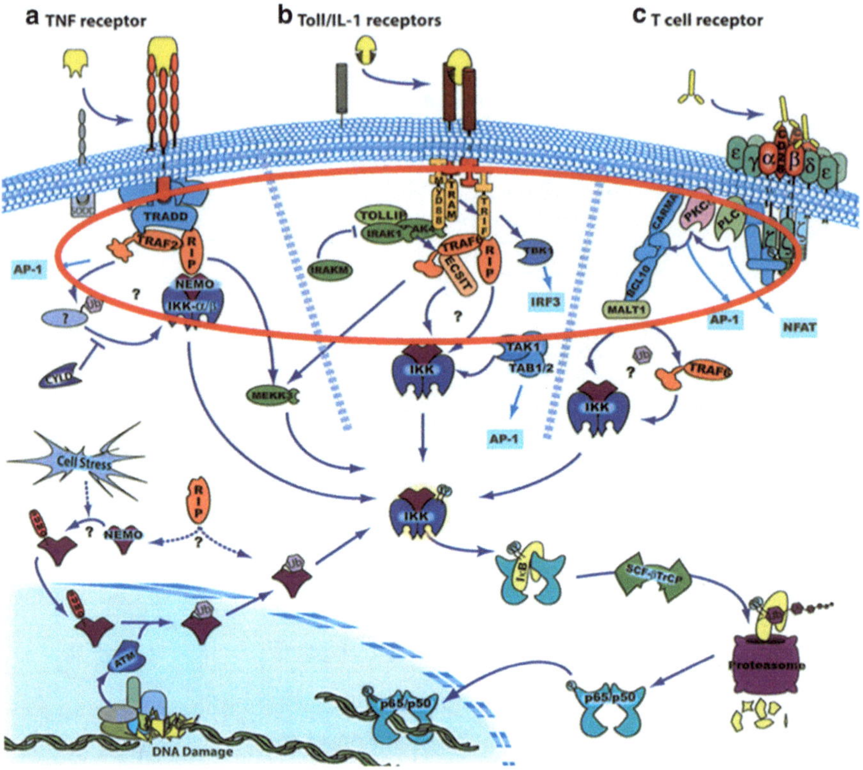

Fig. 1.9 Assembly of adaptor proteins in major signaling pathways that lead to NF-κB activation (Reprinted from Hayden MS, Ghosh S (2004) Signaling to NF-kappaB. *Genes Dev* 18: 2195-2224)

The dimerization of IKK is mediated by NEMO, which can form higher-order structures in vitro (dimers, trimers, and tetramers) and in vivo [143–145]. RIP1 was shown to have a major role in inducing oligomerization of NEMO [113], and activating IKK [112]. Mutations in the NEMO oligomerization domain can prevent IKK function. Moreover, mutated NEMO acts as a dominant negative [144, 146]. Additional proof of the role of IKK oligomerization in induction of NF-κB comes from the observation that viral proteins that activate NF-κB also proceed via oligomerization of NEMO [147, 148]. Thus, oligomerization of NEMO is essential for inducible assembly and activation of the IKK complex.

How does IKK/NEMO oligomerize following signal induction? There are two models that could explain the formation of oligomeric signaling complexes that recruit the IKK subunits and thereby facilitates IKK activation [104].

According to the first model, receptor engagement causes upstream adapter proteins like the RIPs, TRAFs, or various CARD-containing proteins to oligomerize and form higher-order structures. IKK is now recruited and binds to these adapter proteins. This causes IKK oligomerization. It has been reported that oligomerization of RIP or BCL10 and consequent binding of NEMO would recruit the IKK complex to the receptor [113].

According to a second model, ubiquitin chains provide the oligomeric platform and recruit the TAK1 and/or IKK complexes, thereby activating them. There are two conceivable ways in which these ubiquitin chains may act.

The ubiquitin chains may provide the oligomeric platform for another kinase to phosphorylate the proximal IKK or the chains may assist proximity-induced *trans-autophosphorylation*. Alternatively, ubiquitin chains may directly activate the IKK complex. The nature of the linkage of these ubiquitin chains has been proposed to be quite diverse. They may be K63-linked or linear. Traditionally, the ubiquitin chains have been known to be attached to a substrate. However, in addition to these traditional ubiquitin chains, free ubiquitin chains (K63-linked) not attached to any substrate have also been proposed to activate the IKK complex. Interestingly, while K63-linked polyubiquitin chains have been proposed to play a non-degradative role in NF-κB activation, in vitro experiments reveal that proteins modified with K63-linked polyubiquitin chains are actually degraded by the proteasome [149]. However, quantitative mass-spectrometry suggests that following proteasomal inhibition, K48-linked but not K63-linked polyubiquitin chains accumulate in yeast cells [150].

In fact, one area in the NF-κB signaling field that has attracted much attention recently is the area of "regulatory" ubiquitination (or signaling mediated by K63-lined and linear ubiquitin chains). Initial work showed that IKK could be activated by polyubiquitination via a proteasome-independent mechanism (thus, these polyubiquitin chains were not K48-linked) [75]. However, this publication was met with considerable skepticism as it was difficult to appreciate how these unique polyubiquitin chains would respond to a physiological stimulus. Besides, the claim was based only on in vitro biochemical experiments [151], and it was not clear if E2 or E3 enzymes existed that could make non-K48-linked ubiquitin chains. Then, however, TRAF6 was characterized as an E3 ligase while Ubc13/Uev1A was identified to be the E2 complex responsible for catalyzing the formation of K63-linked ubiquitin chains and inducing IKK activation in vitro [152]. The conclusions were partly supported when a homologous Ubc13/Uev1A E2 complex was found in yeast and was shown to be involved in DNA repair by catalyzing the formation of K63 polyubiquitin chains [153]. This jump-started the work in the field of regulatory ubiquitination. Interestingly though, mice deficient in TRAF6 develop osteoporosis, but do not show any drastic phenotype as would be expected of mice lacking any of the crucial components of the NF-κB signaling pathway [154, 155].

Thus, mechanistically speaking IKK could be activated either by oligomerization of adapter proteins or by ubiquitin scaffolds. However, these two mechanisms may not be mutually exclusive. Indeed, proteins that play a role in assembling the signaling complexes (as adaptors) may also get ubiquitinated (like RIP1 or TRAFs). This makes it difficult to ascertain whether the ubiquitin-independent adaptor property of these proteins or the ubiquitin chains on the adaptor proteins is responsible for downstream IKK activation. Examples of adaptor proteins that get ubiquitinated include RIP1, NEMO, and TRAF6, as discussed below.

RIP1 oligomerization leads to inducible interaction with NEMO and subsequent activation of the IKK complex [113]. However, RIP1 is also ubiquitinated at Lys-377 after TNFα stimulation and it has been proposed that it is through these ubiquitin chains that RIP1 binds to NEMO, thus recruiting IKK to the receptor complex (via the UBD of NEMO) [156, 157]. The authors show that the K377R mutant cannot be ubiquitinated and prevents the recruitment of IKK complex to the TNF receptor. They claim that this observation demonstrates the importance of ubiquitin chains in the recruitment process. Their conclusion is however compromised by the fact that the same report reveals that K377R-RIP1 is structurally altered and cannot itself be recruited to the receptor.

NEMO oligomerization can lead to NF-κB activation [143–145]. NEMO has also been reported to be ubiquitinated in several pathways at K399 as well. However, mutation of this site did not lead to complete NF-κB inactivation [158]. Additionally, a knock-in mouse containing a mutation at this site was found to be largely normal [159]. NEMO also contains a UBD and could interact with ubiquitin chains [122]. However, it is difficult to accurately determine the domain required for interaction as the NEMO oligomerization domain overlaps with the NEMO K63-UBD.

Another protein whose autoubiquitination might lead to the activation of a possible IKK-activating kinase (TAK1) is TRAF6. TAK1 has, in turn, been proposed to be activated by TRAF6-catalyzed K63-linked ubiquitin chains [160]. TAK1 exists in a complex with TAB1 and TAB2. TAB2 and TAB3 have UBDs, which bind the K63 chains. An elegant mechanism was proposed in which the TAK1–TAB2–TAB3 complex bound to K63-polyubiquitin chains via the UBD of TAB2 and TAB3, resulting in activation of TAK1. TAK1 subsequently phosphorylated IKK in the two serines in the activation loop, resulting in IKK activation. Another report by the same group subsequently showed that TAK1 could also be activated by unanchored K63-linked polyubiquitin chains (these chains are not linked to any substrate but are free) [161]. From their findings, the authors proposed that K63 ubiquitin chains directly [162] regulate NF-κB activation. However, their model is not substantiated by genetic evidence. Analysis of mice deficient in TAK1, TAB2, and TAB3 shows that while TAK1 plays an essential role in vivo, TAB2 and TAB3 are dispensable. While TAK1 knockouts exhibit embryonic lethality, TAB2 knockouts and TAB3 knockouts show normal NF-κB activation [162]. It is important to note that the proposed dependence on regulatory ubiquitination comes from the binding of TAB2/TAB3 to the K63 liked ubiquitin chains. This binding is important as TAK1 does not itself have a UBD. In theory, the data with the TAB2 and TAB3 knockouts suggests that TAK1 could also be activated by recruitment to the receptor complex by various adaptor proteins; such an alternative model would preclude the requirement of the K63 ubiquitin chains. Of course, it is possible that the TAB2 is compensating for TAB3 and vice versa, but double-knockouts of TAB2 and TAB3 have not yet been reported.

Many other components of the NF-κB signaling pathway have been showed to be modified by K63-linked ubiquitin chains [163]. However, what is not understood is the role these chains play in signaling. One issue with the theory of regulatory ubiquitination is with the delayed kinetics of ubiquitination, and the consistent observation that regulatory ubiquitination only targets a minuscule fraction of any protein. Thus, the hypothesis is that regulatory ubiquitination may result from aggregation of substrates with adaptors exhibiting E3 ligase activity. There are some examples of proteins that are ubiquitinated in a robust manner. But even for these substrates like IRAK1 and MALT1, it is not clear if the regulatory ubiquitination actually precedes IKK activation and IκBα degradation, as would be expected if the K63-linked ubiquitination is indeed "regulatory" [164, 165]. Since it is not clear if these substrates actually get ubiquitinated before IKK activation, there is skepticism as to whether the K63-linked ubiquitin chains are indeed necessary for IKK activation.

Thus, it has been rather challenging to understand the importance of "regulatory ubiquitination" in activating NF-κB. The central problem has been the lack of genetic data. A broad aim of my thesis is trying to decipher the relevance of K63-linked regulatory ubiquitin chains in activation of NF-κB.

1.8 Negative Regulation of NF-κB Signaling

Most of the pathogen-sensing receptors of the innate immune system activate NF-κB and both the innate and adaptive immune system is critically dependent on NF-κB. The importance of this transcription factor can be gauged from the range of receptors that engage the pathway (receptors for proinflammatory cytokines and PAMPs, antigen receptors on T cells and B cells). Thus, the prompt activation of this transcription factor is essential for host defense. Activated NF-κB translocates to the nucleus and induces the transcription of genes responsible for survival, differentiation, and proliferation of immune cells, as well as proinflammatory cytokines that provide the initial protective response against pathogens. However, since the pathway leads to the expression of multiple inflammatory genes, it is critical that the activation is transient and tightly controlled after the initial protective response, as persistent inflammation can lead to tissue damage and other autoimmune diseases. Indeed persistent NF-κB activation is one of the main causes for chronic inflammation and cancer [166–168]. Hence, multiple negative regulatory systems have evolved at different molecular levels to attenuate and terminate signaling [68, 169].

The mammalian NF-κB family consists of several members that have the transactivation domain (TAD) and are thus competent to drive transcription on their own—RelA (p65), RelB, and c-Rel proteins. The family also includes the transcriptionally inactive members like NF-κB1 subunit (p50 derived from p105 precursor) and NF-κB2 subunit (p52 derived from p100 precursor) [24]. p50 and p52 need to form heterodimers with the TAD containing members to be transcriptionally active. The subunits form various homodimers and heterodimers. It is critical that NF-κB is inactive under resting conditions, otherwise the host would be overwhelmed with inflammatory cytokines which have the potential to cause unlimited tissue damage and even death [16].

One method of negative regulation is the synthesis of more inhibitory IκB proteins. Alternatively, since the pathway converges on IKK, turning off upstream signaling

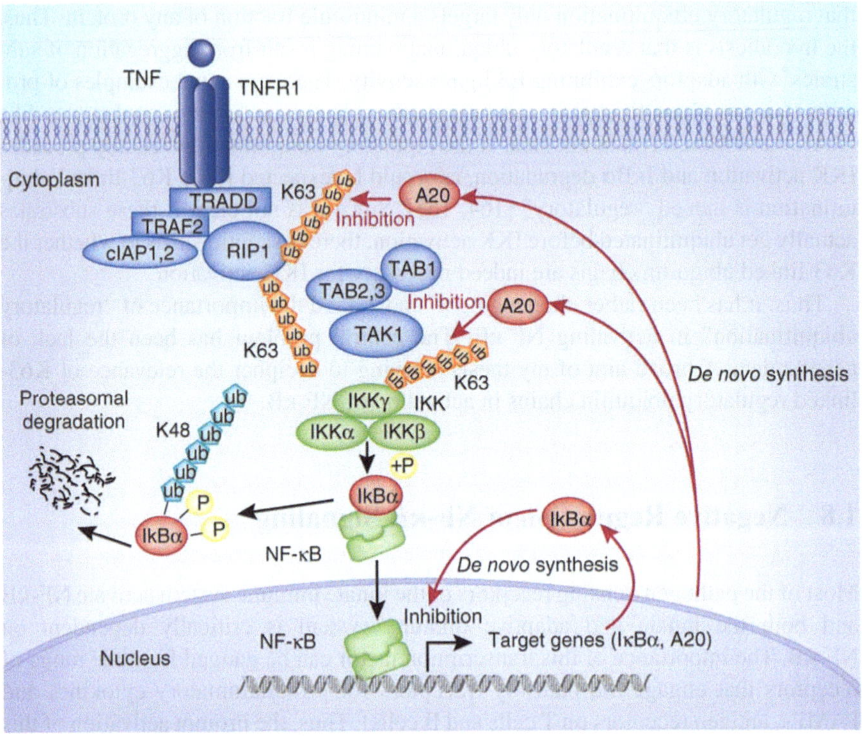

Fig. 1.10 Inhibition of canonical activation of NF-κB by IκBα and A20 (Reprinted by permission from Macmillan Publishers Ltd: [Nature Immunology] (Ruland J (2011) Return to homeostasis: downregulation of NF-[kappa]B responses. *Nat Immunol* 12: 709-714), copyright (2011))

molecules that activate IKK is another method of negative regulation of the pathway [106, 110]. As shown in Fig. 1.10, activation of NF-κB induces expression of negative regulators IκBα and A20. The newly synthesized IκBα protein binds to the NF-κB dimers in the nucleus and shuttle them back into the cytoplasm. The newly synthesized A20 protein acts upstream of IKK by disassembling the IKK-activating complexes.

1.9 Inhibition of NF-κB Activity by Classical IκB Proteins

An overwhelming amount of data points to the biological significance of IκBs as the main negative regulators in NF-κB signaling.

Upon stimulation, a cascade of events activates IKK, which then phosphorylates the inhibitory IκB proteins (here, we will consider only IκBα which serves as a general example) at specific serine residues. IκBα is now proteasomally degraded, leading to the activation of NF-κB. Activated NF-κB induces the expression of

IκBα. The newly synthesized IκBα goes to the nucleus and associates with the NF-κB dimers which are located in the nucleus [170]. IκBα has a nuclear-export sequence, hence the IκB-NF-κB translocates from the nucleus to the cytoplasm. This constitutes a classic negative feedback loop.

Deletion of the main inhibitor of NF-κB (IκBα) expectedly results in early lethality 7–10 days after birth. The cause of death is hyperinflammation, highlighting the importance of NF-κB in the expression of inflammatory genes [72]. Further evidence of IκBα inhibition is provided by the observation that mice which have a mutated IκBα promoter (mutated κB enhancers of the IκBα gene) and consequently produce less IκBα, have a shortened life span (13–15 months), are more sensitive to LPS-mediated septic shock and exhibit abnormal development and activation of T cells [171].

Traditionally, the IκB family of proteins has been thought of as sequestering agents of various NF-κB complexes in the cytoplasm [172]. However, it is now clear that the functions of the individual IκBs do not quite fit into this simplistic criterion. It is probably better to think of the IκB family of proteins as chaperones/cofactors of NF-κB, associating with and stabilizing the NF-κB dimers both in the cytoplasm and nucleus. The binding of IκB to DNA-bound NF-κB potentially influences the recruitment of other coactivators and the resulting transcriptional response. Thus, association of IκB with NF-κB dimers may lead to an increase or decrease of transcription, depending on the context.

An example of an IκB that serves to both activate and inhibit inflammatory gene expression is IκBβ [173]. While IκB generally inhibits NF-κB activation, IκBβ has been found to prolong the expression of certain genes like TNFα or interleukin 1β (IL-1β) [173, 174] as it serves as an essential cofactor for the expression of these genes. It might be counterintuitive to think that the same proteins can perform two opposite functions. However, this is possible as IκBβ exists in two forms that perform opposite functions: a basal phosphorylated form and a hypophosphorylated form that appears upon stimulation.

When cells are stimulated with LPS, the basally phosphorylated IκBβ is slowly degraded (as compared to IκBα). After this, IκBβ is resynthesized as a hypophosphorylated form and can be found in the nucleus [33, 34, 175, 176] with p65-cRel dimers. Thus unlike IκBα, which can dislodge the NF-κB dimers from the DNA, IκBβ cannot promote the spontaneous dissociation of NF-κB dimers from DNA [177]. The two inhibitors are structurally different as IκBβ does not have a NES. Hence, the IκBβ-p65-cRel trimeric complex does not readily export to the cytoplasm. Additionally, the hypophosphorylated form does not mask the NLS of p65. Hence, the trimeric complex is found bound to the DNA [34, 175, 178]. This binding of the hypophosphorylated form of IκBβ to the DNA in association with NF-κB dimers is also supported by the crystal structure of IκBβ bound to homodimers of p65 [179]. In contrast, the phosphorylated IκBβ can mask the p65 NLS, thereby inhibiting binding to DNA under basal conditions [34, 175, 178].

The NF-κB:IκBβ complexes bound to the DNA are stable and resistant to newly synthesized IκBα (and IκBε). Hence, it was hypothesized that nuclear, hypophosphorylated IκBβ may actually be a transcriptional activator of specific genes [34].

This was directly in contrast to the traditional thinking of IκBs only role as cytoplasmic inhibitors of NF-κB.

After nearly a decade, the genetic evidence has substantiated the hypothesis. The IκBβ knockout (made in our lab) indeed shows that IκBβ has distinct functions in the cytoplasm and nucleus [173, 174]. In the cytoplasm, phosphorylated IκBβ acts as a traditional NF-κB inhibitor sequestering p65-cRel heterodimers. However, upon stimulation the cytoplasmic IκBβ is degraded and the heterodimers translocate to the nucleus. Hypophosphorylated IκBβ is newly synthesized and found in the nucleus bound to the p65-c-Rel heterodimers at specific κB promoter sites, accentuating expression of certain proinflammatory genes like TNF [173] and IL-1β [174]. Mice deficient in IκBβ consequently express less TNFα (or IL-1β), rendering them resistant to LPS-induced septic shock and collagen-induced arthritis.

Thus, both genetic and biochemical experiments have demonstrated that IκBβ behaves substantially differently as compared to the other IκB inhibitors.

1.10 Inhibition of NF-κB Activity by Deubiquitinases Upstream of IKK

A plethora of biochemical work, in conjunction with mass-spectrometry, has demonstrated that multiple adaptors upstream of IKK are decorated with K63-linked polyubiquitin chains. Hence, it has been suggested that K63-linked polyubiquitin chains are essential for the activation of NF-κB. The proposition holds that ligand binding to different receptors leads to the recruitment of E3 ligases that polymerize non-degradative K63-linked polyubiquitin chains on specific adaptors [24, 119]. This leads to the activation of IKK in multiple pathways.

For example, stimulation of the TNF receptor results in recruitment of the TRADD adaptor protein, and the consequent assembly of a signalosome consisting of RIP1 and multiple E3 ubiquitin ligases like cIAP1, cIAP2, and TRAF2. This leads to the polyubiquitination of RIP1 by K63-linked polyubiquitin chains [180]. These polyubiquitin chains on RIP1 might play a role in recruiting kinases like TAK1 and IKK (via the UBD of NEMO) [156, 157].

It has been conclusively demonstrated that activation of other receptors also recruits intermediates that have E3 ligase activity. For example, TLRs recognize PAMPs and activate IKK by using various intermediates including MyD88, TRIF, TAK1, and E3 ligases like TRAF6 and IRAK1 [181]. TCR-mediated signaling, on the other hand, proceeds via the recruitment of adaptor proteins like CARD11, Bcl-10, MALT1, TAK1 along with E3 ligases like TRAF2 and TRAF6 [182]. However, there is no genetic evidence to show that the ligase activity is integral to NF-κB activation.

Ubiquitination is a reversible, posttranslational modification, thus deubiquitinases could disassemble the ubiquitin chains and potentially downregulate NF-κB activation. Indeed, deubiquitinases (DUBs) have been shown to be important for the downregulation of NF-κB [114]. It is important to note that by a process known as "ubiquitin editing," DUBs can potentially collaborate with E3 ligases to initially

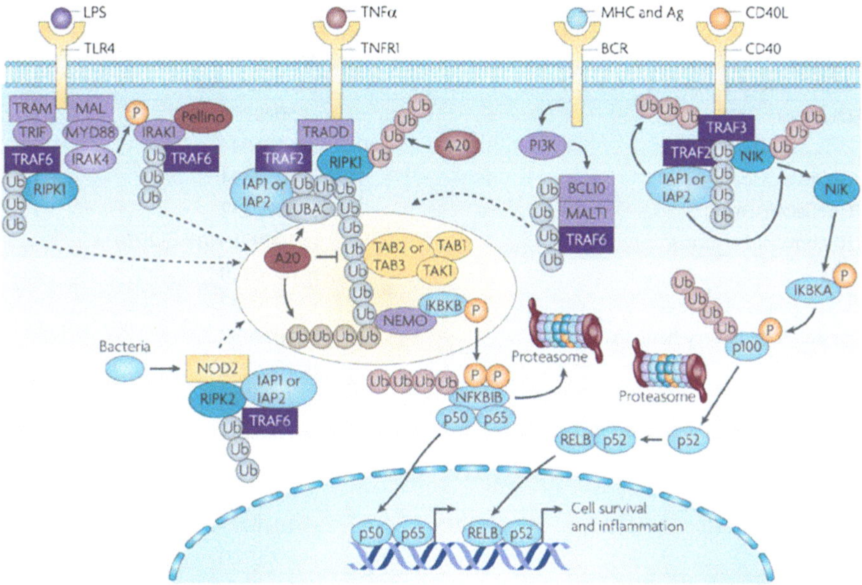

Fig. 1.11 Activation of NF-κB by regulatory ubiquitination and inhibition by deubiquitinases (Reprinted by permission from Macmillan Publishers Ltd: [Nature Reviews Cancer] (Hymowitz SG, Wertz IE (2010) A20: from ubiquitin editing to tumor suppression. *Nat Rev Cancer* 10: 332-341), copyright (2010))

inactivate and then degrade essential mediators of IKK activation. Thus, the DUB could initially deubiquitinate the K63-linked polyubiquitin chains, thereby disassembling the ubiquitin scaffold that organizes the signaling complex. After this, the ligase could add K48-linked polyubiquitin chains to the same protein thereby targeting it for proteasomal degradation [183–185].

As shown in Fig. 1.11, activation of various pathways leads to recruitment of signaling complexes possibly coordinated by the K63-linked polyubiquitination of RIP1 and TRAF6 (red circles). Linear ubiquitin chains may also be formed by the action of the heterodimeric LUBAC complex (linear ubiquitin chain assembly complex; a ubiquitin ligase that catalyzes formation of linear ubiquitin chains) in response to TNF-mediated signaling. IκBα is degraded upon polyubiquitination by K48-linked polyubiquitin chains (blue). Free, unanchored Lys63-linked polyubiquitin chains may also directly activate IKK. A20 may attenuate the signaling by dislodging these activating ubiquitin chains upstream of IKK.

Three common deubiquitinases are A20, Cezanne and CYLD, and mutations or deletions of these genes lead to inflammation or tumorigenesis, as would be expected if these were to function as NF-κB inhibitors. However, as with the ligases, there is no evidence that the deubiquitinase activity of these enzymes is actually responsible for downregulating NF-κB.

Table 1.1 Polymorphisms of A20 associated with autoimmune diseases

Disease	Mutation in *A20* or SNPs identified
Atherosclerosis in mice	E627A (mouse)
Diabetes	rs 5029930
	rs 610604
Crohn's disease	rs 7753394
Coeliac disease	rs 2327832
Rheumatoid arthritis	rs 10499194/rs 13207033
	rs 6920220
	rs 5029937
Systemic lupus erythematosus	rs 5029939
	rs 10499197
	rs 7749323
	rs 13192841
	F127C (rs 2230926)
	rs 6922466
Type 1 diabetes	rs 10499194
	rs 6920220
Psoriasis	rs 610604

Adapted from Trends in Immunology, 30(8), Vereecke L, Beyaert R, van Loo G, The ubiquitin-editing enzyme A20 (TNFAIP3) is a central regulator of immunopathology, 383-391, Copyright (2009), with permission from Elsevier

(1) *A20* (*TNFAIP3*): The best-known deubiquitinase is A20. A20-knockouts die from unrestrained inflammation [127, 186]. Multiple autoimmune diseases like systemic lupus erythematosis are associated with polymorphisms in the A20 locus (Table 1.1). A20 also acts as a tumor suppressor in B-cell lymphoma. Recent evidence also implicates dysfunction of A20 as a risk factor for multiple autoimmune diseases [128, 129, 131, 132, 187]. Thus, A20 is essential for terminating inflammatory responses. Hence, it is very important to understand the mechanism by which A20 functions. A20 is virtually absent under resting conditions and is then rapidly induced by NF-κB [188]. In fact, A20 is also called TNFAIP3 or "TNFα induced protein 3" as it was initially identified following induction with TNFα [189]. Expression of A20 mRNA is under the direct control of NF-κB (much like IκBα). Thus upon stimulation, A20 rapidly accumulates and negatively regulates NF-κB by potentially deubiquitinating K63 chains upstream of IKK. Much like IκBα, this constitutes a classic negative feedback loop.

A20 has a DUB domain followed by a number of C2–C2 zinc-finger domains and the fourth zinc-finger domain (ZF4) is known to have E3 ubiquitin ligase activity [183, 190]. This should, in theory, allow dual ubiquitin-editing function, i.e., A20 should be able to act both as a ligase and a deubiquitinase. It has been biochemically demonstrated that upon stimulation with TNFα, A20 accumulates and disassembles the K63-linked polyubiquitin chains from RIP1, attenuating NF-κB signaling (the

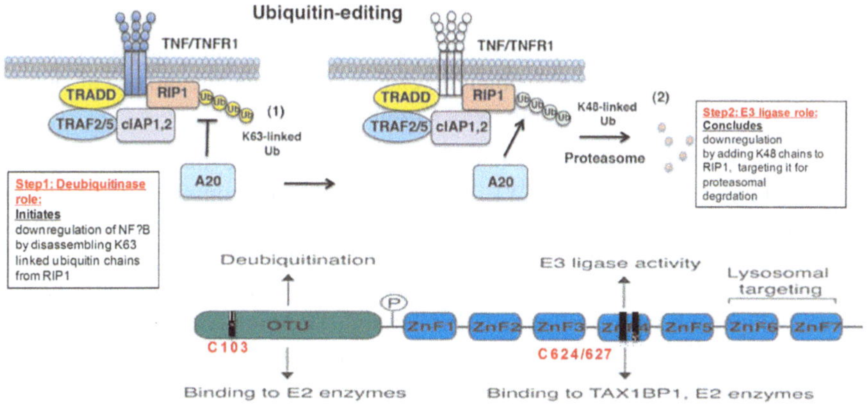

Fig. 1.12 Mechanism of action of A20 (modified with permission from Macmillan Publishers Ltd: [Cell Research] (Harhaj EW, Dixit VM (2011) Deubiquitinases in the regulation of NF-[kappa] B signaling. *Cell Res* 21: 22-39), copyright (2011))

deubiquitinase function is mediated by Cys103 in the OTU domain). In the next step, the E3 ligase domain adds degradative K48-linked polyubiquitination chains to RIP1, leading to proteasomal degradation of RIP1 and terminating NF-κB signaling [183]. Incidentally, the DUB activity of A20 is not specific to RIP1 as A20 generally deubiquitinates other substrates upstream of IKK. These include TRAF6 [186], NEMO [191], RIP2 [192], and MALT1 [193] (Fig. 1.12).

The deubiquitinase activity of A20 has been proposed to be essential in downregulating NF-κB on the basis of in vitro work. However, certain critical biochemical data are also difficult to reconcile with this. For example, overexpression of a deubiquitinase mutant of A20 has been shown to inhibit NF-κB [190, 194]. It is also paradoxical in the light of the proposed mechanism that the deubiquitinase activity has been found to be nonspecific for K63 chains. Indeed A20 has been shown to only weakly deubiquitinate K63 polyubiquitin chains, while it robustly disassembles K48 polyubiquitin chains [195, 196].

Recently, the crystal structure of the OTU domain of A20 has been solved [195, 196] and it is not clear how A20 would distinguish between K48 and K63-linked chains. A20 reacts with polyubiquitinated TRAF6 to yield polyubiquitin chains but not monoubiquitin linkages [196]. This shows that A20 might derive specificity from directly interacting with TRAF6. However, A20 does not cleave monoubiquitinated TRAF6. Thus understanding the mechanism of action and specificity of A20 will require additional studies.

(2) *Cezanne*: Along with A20, there is another cysteine-protease DUB called Cezanne (cellular zinc-finger anti-NF-κB) which also attenuates NF-κB signaling by deubiquitinating K63-linked chains from RIP1 [197]. Like A20, the expression of Cezanne increases after TNFα stimulation. The deubiquitinase action of Cezanne

may be biochemically important as the catalytically inactive mutant cannot inhibit NF-κB activation. However Cezanne is not specific to K63 chains and preferentially deubiquitinates K11-linked polyubiquitin chain [198]. Thus, the biological significance of the deubiquitinating activity of Cezanne is unclear.

(3) *CYLD*: The third well-known deubiquitinase is CYLD (cylindromatosis gene). The deubiquitinase action of CYLD is mediated by the conserved Cys601 and mutating this residue leads to persistent NF-κB signaling [199]. It has been biochemically established that CYLD negatively regulates NF-κB [133, 199, 200]. However, CYLD is a promiscuous deubiquitinase as it removes ubiquitin chains from multiple substrates including RIP1, TRAF2, TRAF6, TAK1, and IKKγ [133, 199, 201–203]. In this regard, CYLD lacks specificity and cleaves both K63-linked polyubiquitin chains as well as linear ubiquitin chains [204, 205]. It can also cleave K48-linked chains from adaptor Lck in T cells [206].

The expression of CYLD is not directly induced by NF-κB (unlike A20 and Cezanne) [207]. Thus, CYLD is present at basal levels and might play a role in suppressing NF-κB activation in resting cells. In contrast, A20 is expressed at low levels under basal conditions and rapidly induced in response to proinflammatory stimuli to downregulate excessive NF-κB activation. This may be the reason why the two DUBs do not compensate for each other as their mechanisms of action are temporally distinct. The activity of CYLD has to be depressed for the cells to become activated. Hence, its activity is controlled by subcellular localization in addition to phosphorylation on multiple serine residues by IKK (in a NEMO-dependent manner) [208, 209].

Since CYLD deubiquitinates multiple upstream activators of IKK, the expectation was that the CYLD knockout would show a drastic phenotype. Accordingly, multiple groups embarked on the project and mice deficient in CYLD were generated independently by various groups. Contrary to expectations, *Cyld*$^{-/-}$ mice show no obvious defect in growth or survival [206]. However, they do develop some defects in the immune system with age. Additionally, the knockouts generated by different groups sometimes have conflicting phenotypes. In general, these genetic knockouts show that CYLD plays a role in immune cell development, osteoclastogenesis, and spermatogenesis. Thus, *Cyld*$^{-/-}$ mice made by one group have reduced numbers of CD4(+) and CD8(+) single positive T cells in the thymus and periphery [206]. *Cyld*$^{-/-}$ mice spontaneously develop inflammation in the colon [203], osteoporosis [210], B-cell hyperplasia, and lymphoid organ enlargement [211]. Studies also show that CYLD plays an important role for survival of immature NKT cells [212] and spermatogenesis [201]. However conflicting genetic data from another group showed that development of T cells and myeloid cells was normal in CYLD-knockouts [202]. A third group showed that Cyld-deficient mice are more prone to skin tumors [208]. Mice expressing a naturally occurring variant of CYLD that did not express exon 7 and 8 (thereby lacking binding sites to TRAF2 and NEMO) were found to have enlarged spleens as a result of dramatic accumulation of mature B cells [213].

1.11 Perspectives on Inhibitors of NF-κB and Broad Aims of this Thesis

Twenty-five years of research has made it clear that appropriate downregulation of NF-κB signaling is as important as NF-κB activation. IκB proteins are the most well-documented inhibitors of NF-κB in the physiological setting, as borne out by both biochemical and genetic studies. However along with the IκBs, there is a plethora of biochemical evidence to suggest that deubiquitinases may also function as negative feedback regulators of NF-κB signaling. This thesis examines the mechanism of one major inhibitor of NF-κB: A20, which downregulates activation of NF-κB upstream of IKK in stimulated cells.

Broad aim: *Examine the role of regulatory K63-linked polyubiquitin chains in NF-κB activation (specifically, I examined the role of deubiquitinase activity of A20 by generating knock-in mice selectively lacking the K63-deubiquitinase function).*

While multiple adaptor proteins have been shown to be covalently linked with K63-linked polyubiquitin chains, it is not clear if such regulatory polyubiquitination is essential for activation of IKK in the physiological context. This is because of an acute lack of genetic complementation experiments. Ultimately, the importance of regulatory ubiquitination mediated by K63-linked polyubiquitin chains has to be demonstrated in knock-in mouse models (that have point mutations selectively disrupting regulatory ubiquitination). The need for this genetic evidence can be understood by studies done on NEMO. NEMO knockouts are embryonically lethal, thus demonstrating that NEMO is clearly essential for NF-κB activation [69, 214]. Biochemical experiments showed that K63-linked polyubiquitination of Lys-399 was essential for NF-κB activation [158], and it was believed for some time that this ubiquitination would play an important role in animals. However, a knock-in mouse with a point mutation at this site was found to be largely normal with respect to NF-κB activation [159]. This shows that K63-linked polyubiquitination of Lys-399 is not important for NF-κB activation.

Traditionally, putative targets that have been proposed to modulate NF-κB activation by regulatory ubiquitination were identified by overexpression of proteins mutated at specific lysines. However, mutating a positively charged residue like lysine might change the protein conformation leading to incorrect protein folding, thereby altering other functional attributes of the protein. For example, the ubiquitination of RIP1 at K377 [156, 215] in response to TNFα was demonstrated by overexpression and knockdown studies to be essential in activating IKK. The K377R mutant is not ubiquitinated; however it also does not get recruited to the activated TNF receptor complex [156]. This may be because of altered conformations in the mutated protein. In fact, a subsequent study by the same group has shown that cells which have endogenous ubiquitin genetically replaced with K63R ubiquitin do not show impaired IKK activation in response to TNFα [216]. One lesson to draw from this discrepancy is that knockdown and overexpression studies have limitations. The level of protein depletion affects experimental outcomes in knockdown studies, while overexpression of signaling components such as ubiquitin ligases or adaptor proteins (like the TRAFs and cIAPs) may lead to promiscuous oligomerization and signaling.

In order to evaluate the effect of regulatory ubiquitination on the NF-κB pathway, we examined various genes that are involved in regulatory ubiquitination based on previously published genetic and biochemical data. Specifically, we looked for two criteria:

1. The knockout of this gene should result in a phenotype consistent with a profound dysregulation of NF-κB signaling, thus this gene would have to be non-redundant in terms of its effects on NF-κB signaling.
2. Biochemical and in vitro complementation studies of the gene in cell lines (using different methodologies) would have to clearly point to an amino-acid that is involved in regulatory ubiquitination, such that mutating this residue would disrupt regulatory ubiquitination and NF-κB signaling.

Based on these criteria, we planned to generate a knock-in mouse with a point mutation for the identified residue of interest. If regulatory ubiquitination is indeed important for NF-κB signaling, then this knock-in mouse would be expected to resemble the knockout (at least partially). If not, then some other property of the protein (apart from the ability to mediate regulatory ubiquitination) would be expected to be involved in NF-κB signaling.

The candidate genes that we considered included: E2 enzymes (Ubc13), E3 enzymes (like TRAFs and cIAPs), adaptor proteins that are ubiquitinated by K63-linked ubiquitin chains (like RIP1), and deubiquitinases like A20.

(1) *E2 enzyme Ubc13*: Two groups reported that Ubc13 deficiency causes embryonic lethality; however it is not clear if this lethality results from defective NF-κB signaling [217, 218].

(2) *E3 ligases like TRAF and cIAP proteins*: TRAF proteins are highly redundant. There are a total of seven TRAF proteins, all of which have a common C-terminal coiled-coil TRAF domain. TRAFs2–7 also contains a N-terminal RING finger domain that is involved in ligating ubiquitin chains to substrates; thereby enabling these TRAFs to act as E3 ligases. Amongst them, TRAFs2, 5, and 6 have been shown to be important adaptors for IKK activation. TRAF6 plays a role in LPS-mediated signaling, and TRAF6 knockouts develop osteoporosis [154, 155]. TRAF2 and TRAF5 act downstream of the TNF receptor. TRAF2 knockouts are born normal and do not show any major defect in NF-κB signaling [219]. This might be expected as TRAF5 may compensate for TRAF2. Indeed, TRAF5 knockouts are also relatively normal because of compensation by TRAF2 [220]. However, TRAF2/5 double-knockouts are embryonic lethal [221]. It is not clear though if the RING domain of the TRAFs is essential for NF-κB activation [222]. Additionally, along with the E3 ligase action, the RING domain of TRAF2 is also essential for recruiting IKK to the receptor complex. Hence, it would be difficult to ascertain if any observed signaling defect would

be because of "RING-mediated regulatory ubiquitination" or "RING-mediated adaptor function."

Similar to the TRAFs, cIAP1 and cIAP2 are also redundant and compensate for each other. None of the single-knockouts show a phenotype [223, 224], because of compensation by the other isotype. Cells deficient in both the isotypes are sensitized to apoptosis resulting from decreased activation of NF-κB [180, 225].

(3) *Adaptor proteins like RIP1*: Mice deficient in RIP1 die within a couple of days after birth from tissue apoptosis as the antiapoptotic NF-κB pathway cannot be activated in these animals [226]; thus RIP1 clearly satisfies the "genetic criteria" above. Moreover, RIP1 has been shown to be polyubiquitinated by both K48 and K63 chains using linkage-specific antibodies [184] and by mass-spectrometry [227–229] in response to TNF signaling (mass-spec analysis additionally shows evidence of K11 and linear ubiquitination). K377 of RIP1 was identified as the site modified with K63-linked ubiquitination; however as explained above, the mutated protein (K377R) is inactive as does not get recruited to the receptor complex. Other reports though have shown that RIP1 can also be ubiquitinated at Lys115, Lys570, Lys603, and Lys626 when overexpressed [230], and also at Lys115, Lys163–167, and Lys671 under endogenous conditions [231]. Moreover, there have been conflicting reports which suggest that RIP1 may not be needed for TNF-mediated NF-κB activation [232, 233]. Hence, we decided to forgo RIP1 as the "biochemical criteria" is not clearly satisfied [234].

(4) *Deubiquitinase protein (A20)*: A20 is a deubiquitinase that downregulates NF-κB signaling and completely satisfies both the criteria outlined above. In fact, the genetic and biochemical work done on A20 suggests that K63-linked polyubiquitination may indeed play a role in NF-κB signaling [163]. A20 knockouts die prematurely from rampant inflammation of multiple organs and cachexia as these animals are unable to downregulate persistent NF-κB signaling [127]. Hence, it is clear that A20 is non-redundant in terms of downregulating NF-κB. A20 has been extensively characterized as a protein that acts on both K48- and K63-linked polyubiquitin chains. Multiple biochemical evidences have shown that it may act by first removing K63-linked ubiquitin chains from adapter proteins such as RIP1, thereby attenuating NF-κB signaling. After this, A20 adds K48-linked polyubiquitin chains to RIP1 and targets it for proteasomal degradation, resulting in a complete block of NF-κB-induced signals [183] (Fig. 1.13). The initial deubiquitinase function is mediated by Cys103, and the subsequent E3-ligase function is mediated by Cys624/Cys627. Mutating any of these residues leads to persistent activation of NF-κB in vitro.

Hence, *to broadly ascertain the physiological significance of regulatory K63-linked polyubiquitination, our specific aim is to generate the C103A knock-in mouse and characterize the effects of the deubiquitinase activity of A20*. If Cys103 is indeed the catalytic residue responsible for the function of A20, then these mice would be expected to resemble the inflammatory phenotype of mice deficient in A20 (Fig. 1.13).

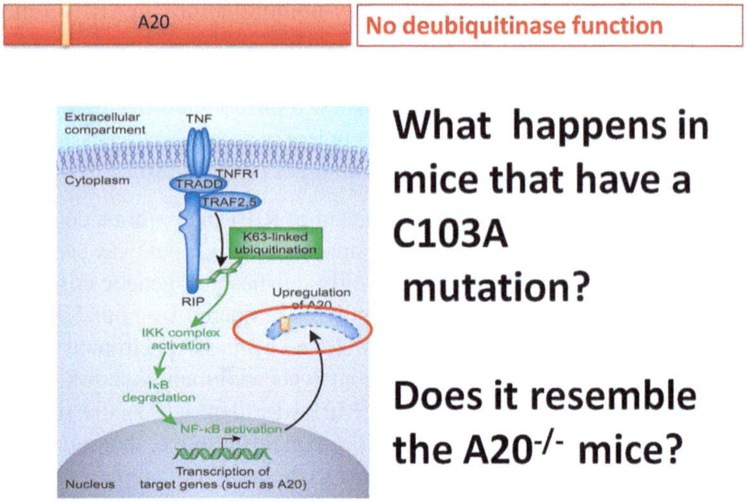

Fig. 1.13 Aim: Determining physiological contribution of deubiquitinase function of A20 (modified with permission from Macmillan Publishers Ltd: [Nature Immunology] (Vigo Heissmeyer and Anjana Rao (2008) Itching to end NF-kappaB. *Nature Immunology* 9(3): 227-229), copyright (2008))

References

1. Dethlefsen L, McFall-Ngai M, Relman DA (2007) An ecological and evolutionary perspective on human-microbe mutualism and disease. Nature 449:811–818
2. Smith DA, Germolec DR (1999) Introduction to immunology and autoimmunity. Environ Health Perspect 107(Suppl 5):661–665
3. Medzhitov R (2001) Toll-like receptors and innate immunity. Nat Rev Immunol 1:135–145
4. Gregersen PK, Behrens TW (2006) Genetics of autoimmune diseases—disorders of immune homeostasis. Nat Rev Genet 7:917–928
5. Iwasaki A, Medzhitov R (2004) Toll-like receptor control of the adaptive immune responses. Nat Immunol 5:987–995
6. Janeway CA Jr (1989) Approaching the asymptote? Evolution and revolution in immunology. Cold Spring Harb Symp Quant Biol 54(Pt 1):1–13
7. Janeway CA Jr, Medzhitov R (2002) Innate immune recognition. Annu Rev Immunol 20:197–216
8. Linehan SA, Martinez-Pomares L, Gordon S (2000) Macrophage lectins in host defence. Microbes Infect 2:279–288
9. Chen G, Shaw MH, Kim YG, Nunez G (2009) NOD-like receptors: role in innate immunity and inflammatory disease. Annu Rev Pathol 4:365–398
10. Franchi L, Warner N, Viani K, Nunez G (2009) Function of Nod-like receptors in microbial recognition and host defense. Immunol Rev 227:106–128
11. Inohara N, Nunez G (2003) NODs: intracellular proteins involved in inflammation and apoptosis. Nat Rev Immunol 3:371–382
12. Yoneyama M, Fujita T (2009) RNA recognition and signal transduction by RIG-I-like receptors. Immunol Rev 227:54–65
13. Johannessen M, Askarian F, Sangvik M, Sollid JE (2013) Bacterial interference with canonical NFkappaB signaling. Microbiology 159(Pt 10):2001–2013. doi:10.1099/mic.0.069369-0

14. Mogensen TH (2009) Pathogen recognition and inflammatory signaling in innate immune defenses. Clin Microbiol Rev 22(2):240–273, Table of Contents
15. Courtois G, Gilmore TD (2006) Mutations in the NF-kappaB signaling pathway: implications for human disease. Oncogene 25:6831–6843
16. Karin M, Greten FR (2005) NF-kappaB: linking inflammation and immunity to cancer development and progression. Nat Rev Immunol 5:749–759
17. Karin M (2006) Nuclear factor-kappaB in cancer development and progression. Nature 441:431–436
18. Ma A, Malynn BA (2012) A20: linking a complex regulator of ubiquitylation to immunity and human disease. Nat Rev Immunol 12:774–785
19. Hayden MS, West AP, Ghosh S (2006) NF-kappaB and the immune response. Oncogene 25:6758–6780
20. Sen R, Baltimore D (1986) Multiple nuclear factors interact with the immunoglobulin enhancer sequences. Cell 46:705–716
21. Sen R, Baltimore D (1986) Inducibility of kappa immunoglobulin enhancer-binding protein Nf-kappa B by a posttranslational mechanism. Cell 47:921–928
22. Baeuerle PA, Baltimore D (1988) Activation of DNA-binding activity in an apparently cytoplasmic precursor of the NF-kappa B transcription factor. Cell 53:211–217
23. Hinz M, Arslan SÇ, Scheidereit C (2012) It takes two to tango: IκBs, the multifunctional partners of NF-κB. Immunol Rev 246:59–76
24. Hayden MS, Ghosh S (2008) Shared principles in NF-kappaB signaling. Cell 132:344–362
25. Hoffmann A, Natoli G, Ghosh G (2006) Transcriptional regulation via the NF-kappaB signaling module. Oncogene 25:6706–6716
26. Perkins ND (2006) Post-translational modifications regulating the activity and function of the nuclear factor kappa B pathway. Oncogene 25:6717–6730
27. Wan F et al (2007) Ribosomal protein S3: a KH domain subunit in NF-kappaB complexes that mediates selective gene regulation. Cell 131:927–939
28. Liou HC, Nolan GP, Ghosh S, Fujita T, Baltimore D (1992) The NF-kappa B p50 precursor, p105, contains an internal I kappa B-like inhibitor that preferentially inhibits p50. EMBO J 11:3003–3009
29. Dobrzanski P, Ryseck RP, Bravo R (1995) Specific inhibition of RelB/p52 transcriptional activity by the C-terminal domain of p100. Oncogene 10:1003–1007
30. Whiteside ST, Israel A (1997) I kappa B proteins: structure, function and regulation. Semin Cancer Biol 8:75–82
31. Weil R, Whiteside ST, Israel A (1997) Control of NF-kappa B activity by the I kappa B beta inhibitor. Immunobiology 198:14–23
32. Karin M, Ben-Neriah Y (2000) Phosphorylation meets ubiquitination: the control of NF-[kappa]B activity. Annu Rev Immunol 18:621–663
33. Thompson JE, Phillips RJ, Erdjument-Bromage H, Tempst P, Ghosh S (1995) I kappa B-beta regulates the persistent response in a biphasic activation of NF-kappa B. Cell 80:573–582
34. Suyang H, Phillips R, Douglas I, Ghosh S (1996) Role of unphosphorylated, newly synthesized I kappa B beta in persistent activation of NF-kappa B. Mol Cell Biol 16:5444–5449
35. Li Z, Nabel GJ (1997) A new member of the I kappaB protein family, I kappaB epsilon, inhibits RelA (p65)-mediated NF-kappaB transcription. Mol Cell Biol 17:6184–6190
36. Whiteside ST, Epinat JC, Rice NR, Israel A (1997) I kappa B epsilon, a novel member of the I kappa B family, controls RelA and cRel NF-kappa B activity. EMBO J 16:1413–1426
37. Ghosh S, Baltimore D (1990) Activation in vitro of NF-kappa B by phosphorylation of its inhibitor I kappa B. Nature 344:678–682
38. Naumann M, Scheidereit C (1994) Activation of NF-kappa B in vivo is regulated by multiple phosphorylations. EMBO J 13:4597–4607
39. Brown K, Gerstberger S, Carlson L, Franzoso G, Siebenlist U (1995) Control of I kappa B-alpha proteolysis by site-specific, signal-induced phosphorylation. Science 267:1485–1488

40. Brockman JA, Scherer DC, McKinsey TA, Hall SM, Qi X, Lee WY, Ballard DW (1995) Coupling of a signal response domain in I kappa B alpha to multiple pathways for NF-kappa B activation. Mol Cell Biol 15:2809–2818
41. Alkalay I, Yaron A, Hatzubai A, Jung S, Avraham A, Gerlitz O, Pashut-Lavon I, Ben-Neriah Y (1995) In vivo stimulation of I kappa B phosphorylation is not sufficient to activate NF-kappa B. Mol Cell Biol 15:1294–1301
42. DiDonato JA, Mercurio F, Karin M (1995) Phosphorylation of I kappa B alpha precedes but is not sufficient for its dissociation from NF-kappa B. Mol Cell Biol 15:1302–1311
43. Finco TS, Beg AA, Baldwin AS Jr (1994) Inducible phosphorylation of I kappa B alpha is not sufficient for its dissociation from NF-kappa B and is inhibited by protease inhibitors. Proc Natl Acad Sci U S A 91:11884–11888
44. Lin R, Beauparlant P, Makris C, Meloche S, Hiscott J (1996) Phosphorylation of IkappaBalpha in the C-terminal PEST domain by casein kinase II affects intrinsic protein stability. Mol Cell Biol 16:1401–1409
45. Miyamoto S, Maki M, Schmitt MJ, Hatanaka M, Verma IM (1994) Tumor necrosis factor alpha-induced phosphorylation of I kappa B alpha is a signal for its degradation but not dissociation from NF-kappa B. Proc Natl Acad Sci U S A 91:12740–12744
46. Beg AA, Finco TS, Nantermet PV, Baldwin AS Jr (1993) Tumor necrosis factor and interleukin-1 lead to phosphorylation and loss of I kappa B alpha: a mechanism for NF-kappa B activation. Mol Cell Biol 13:3301–3310
47. Brown K, Park S, Kanno T, Franzoso G, Siebenlist U (1993) Mutual regulation of the transcriptional activator NF-kappa B and its inhibitor, I kappa B-alpha. Proc Natl Acad Sci U S A 90:2532–2536
48. Mellits KH, Hay RT, Goodbourn S (1993) Proteolytic degradation of MAD3 (I kappa B alpha) and enhanced processing of the NF-kappa B precursor p105 are obligatory steps in the activation of NF-kappa B. Nucleic Acids Res 21:5059–5066
49. Sun SC, Ganchi PA, Ballard DW, Greene WC (1993) NF-kappa B controls expression of inhibitor I kappa B alpha: evidence for an inducible autoregulatory pathway. Science 259:1912–1915
50. Lin YC, Brown K, Siebenlist U (1995) Activation of NF-kappa B requires proteolysis of the inhibitor I kappa B-alpha: signal-induced phosphorylation of I kappa B-alpha alone does not release active NF-kappa B. Proc Natl Acad Sci U S A 92:552–556
51. Henkel T, Machleidt T, Alkalay I, Kronke M, Ben-Neriah Y, Baeuerle PA (1993) Rapid proteolysis of I kappa B-alpha is necessary for activation of transcription factor NF-kappa B. Nature 365:182–185
52. Chen Z, Hagler J, Palombella VJ, Melandri F, Scherer D, Ballard D, Maniatis T (1995) Signal-induced site-specific phosphorylation targets I kappa B alpha to the ubiquitin-proteasome pathway. Genes Dev 9:1586–1597
53. Palombella VJ, Rando OJ, Goldberg AL, Maniatis T (1994) The ubiquitin-proteasome pathway is required for processing the NF-kappa B1 precursor protein and the activation of NF-kappa B. Cell 78:773–785
54. Alkalay I, Yaron A, Hatzubai A, Orian A, Ciechanover A, Ben-Neriah Y (1995) Stimulation-dependent I kappa B alpha phosphorylation marks the NF-kappa B inhibitor for degradation via the ubiquitin-proteasome pathway. Proc Natl Acad Sci U S A 92:10599–10603
55. Spencer E, Jiang J, Chen ZJ (1999) Signal-induced ubiquitination of IkappaBalpha by the F-box protein Slimb/beta-TrCP. Genes Dev 13:284–294
56. Tan P, Fuchs SY, Chen A, Wu K, Gomez C, Ronai Z, Pan ZQ (1999) Recruitment of a ROC1-CUL1 ubiquitin ligase by Skp1 and HOS to catalyze the ubiquitination of I kappa B alpha. Mol Cell 3:527–533
57. Winston JT, Strack P, Beer-Romero P, Chu CY, Elledge SJ, Harper JW (1999) The SCFbeta-TRCP-ubiquitin ligase complex associates specifically with phosphorylated destruction motifs in IkappaBalpha and beta-catenin and stimulates IkappaBalpha ubiquitination in vitro. Genes Dev 13:270–283
58. Yaron A et al (1998) Identification of the receptor component of the IkappaBalpha-ubiquitin ligase. Nature 396:590–594

59. Hoffmann A, Levchenko A, Scott ML, Baltimore D (2002) The IkappaB-NF-kappaB signaling module: temporal control and selective gene activation. Science 298:1241–1245
60. Chen L, Fischle W, Verdin E, Greene WC (2001) Duration of nuclear NF-kappaB action regulated by reversible acetylation. Science 293:1653–1657
61. Arenzana-Seisdedos F, Turpin P, Rodriguez M, Thomas D, Hay RT, Virelizier JL, Dargemont C (1997) Nuclear localization of I kappa B alpha promotes active transport of NF-kappa B from the nucleus to the cytoplasm. J Cell Sci 110(Pt 3):369–378
62. Zandi E, Rothwarf DM, Delhase M, Hayakawa M, Karin M (1997) The IkappaB kinase complex (IKK) contains two kinase subunits, IKKalpha and IKKbeta, necessary for IkappaB phosphorylation and NF-kappaB activation. Cell 91:243–252
63. Mercurio F et al (1997) IKK-1 and IKK-2: cytokine-activated IkappaB kinases essential for NF-kappaB activation. Science 278:860–866
64. Rothwarf DM, Zandi E, Natoli G, Karin M (1998) IKK-gamma is an essential regulatory subunit of the IkappaB kinase complex. Nature 395:297–300
65. Woronicz JD, Gao X, Cao Z, Rothe M, Goeddel DV (1997) IkappaB kinase-beta: NF-kappaB activation and complex formation with IkappaB kinase-alpha and NIK. Science 278:866–869
66. Yamaoka S, Courtois G, Bessia C, Whiteside ST, Weil R, Agou F, Kirk HE, Kay RJ, Israel A (1998) Complementation cloning of NEMO, a component of the IkappaB kinase complex essential for NF-kappaB activation. Cell 93:1231–1240
67. Lawrence T, Bebien M, Liu GY, Nizet V, Karin M (2005) IKKalpha limits macrophage NF-kappaB activation and contributes to the resolution of inflammation. Nature 434:1138–1143
68. Shih VF, Tsui R, Caldwell A, Hoffmann A (2011) A single NFkappaB system for both canonical and non-canonical signaling. Cell Res 21:86–102
69. Rudolph D, Yeh W-C, Wakeham A, Rudolph B, Nallainathan D, Potter J, Elia AJ, Mak TW (2000) Severe liver degeneration and lack of NF-κB activation in NEMO/IKKγ-deficient mice. Genes Dev 14:854–862
70. Delhase M, Hayakawa M, Chen Y, Karin M (1999) Positive and negative regulation of IkappaB kinase activity through IKKbeta subunit phosphorylation. Science 284:309–313
71. Li Q, Antwerp DV, Mercurio F, Lee K-F, Verma IM (1999) Severe Liver Degeneration in Mice Lacking the IκB Kinase 2 Gene. Science 284:321–325
72. Beg AA, Sha WC, Bronson RT, Ghosh S, Baltimore D (1995) Embryonic lethality and liver degeneration in mice lacking the RelA component of NF-kappa B. Nature 376:167–170
73. Huang TT, Kudo N, Yoshida M, Miyamoto S (2000) A nuclear export signal in the N-terminal regulatory domain of IkappaBalpha controls cytoplasmic localization of inactive NF-kappaB/IkappaBalpha complexes. Proc Natl Acad Sci U S A 97:1014–1019
74. Johnson C, Van Antwerp D, Hope TJ (1999) An N-terminal nuclear export signal is required for the nucleocytoplasmic shuttling of IkappaBalpha. EMBO J 18:6682–6693
75. Chen ZJ, Parent L, Maniatis T (1996) Site-specific phosphorylation of IkappaBalpha by a novel ubiquitination-dependent protein kinase activity. Cell 84:853–862
76. Scheidereit C (2006) IkappaB kinase complexes: gateways to NF-kappaB activation and transcription. Oncogene 25:6685–6705
77. Ghosh S, Karin M (2002) Missing pieces in the NF-kappaB puzzle. Cell 109(Suppl):S81–S96
78. Bonizzi G, Karin M (2004) The two NF-kappaB activation pathways and their role in innate and adaptive immunity. Trends Immunol 25:280–288
79. Novack DV, Yin L, Hagen-Stapleton A, Schreiber RD, Goeddel DV, Ross FP, Teitelbaum SL (2003) The IkappaB function of NF-kappaB2 p100 controls stimulated osteoclastogenesis. J Exp Med 198:771–781
80. Saitoh T, Nakayama M, Nakano H, Yagita H, Yamamoto N, Yamaoka S (2003) TWEAK induces NF-kappaB2 p100 processing and long lasting NF-kappaB activation. J Biol Chem 278:36005–36012
81. Claudio E, Brown K, Park S, Wang H, Siebenlist U (2002) BAFF-induced NEMO-independent processing of NF-kappa B2 in maturing B cells. Nat Immunol 3:958–965

82. Coope HJ, Atkinson PG, Huhse B, Belich M, Janzen J, Holman MJ, Klaus GG, Johnston LH, Ley SC (2002) CD40 regulates the processing of NF-kappaB2 p100 to p52. EMBO J 21:5375–5385
83. Fusco AJ, Savinova OV, Talwar R, Kearns JD, Hoffmann A, Ghosh G (2008) Stabilization of RelB requires multidomain interactions with p100/p52. J Biol Chem 283:12324–12332
84. Basak S, Shih VF, Hoffmann A (2008) Generation and activation of multiple dimeric transcription factors within the NF-kappaB signaling system. Mol Cell Biol 28:3139–3150
85. Yilmaz ZB, Weih DS, Sivakumar V, Weih F (2003) RelB is required for Peyer's patch development: differential regulation of p52-RelB by lymphotoxin and TNF. EMBO J 22:121–130
86. Senftleben U et al (2001) Activation by IKKalpha of a second, evolutionary conserved, NF-kappa B signaling pathway. Science 293:1495–1499
87. Xiao G, Harhaj EW, Sun SC (2001) NF-kappaB-inducing kinase regulates the processing of NF-kappaB2 p100. Mol Cell 7:401–409
88. Liang C, Zhang M, Sun SC (2006) beta-TrCP binding and processing of NF-kappaB2/p100 involve its phosphorylation at serines 866 and 870. Cell Signal 18:1309–1317
89. Amir RE, Haecker H, Karin M, Ciechanover A (2004) Mechanism of processing of the NF-kappa B2 p100 precursor: identification of the specific polyubiquitin chain-anchoring lysine residue and analysis of the role of NEDD8-modification on the SCF(beta-TrCP) ubiquitin ligase. Oncogene 23:2540–2547
90. Xiao G, Fong A, Sun SC (2004) Induction of p100 processing by NF-kappaB-inducing kinase involves docking IkappaB kinase alpha (IKKalpha) to p100 and IKKalpha-mediated phosphorylation. J Biol Chem 279:30099–30105
91. Yin L, Wu L, Wesche H, Arthur CD, White JM, Goeddel DV, Schreiber RD (2001) Defective lymphotoxin-beta receptor-induced NF-kappaB transcriptional activity in NIK-deficient mice. Science 291:2162–2165
92. Fagarasan S, Shinkura R, Kamata T, Nogaki F, Ikuta K, Tashiro K, Honjo T (2000) Alymphoplasia (aly)-type nuclear factor kappaB-inducing kinase (NIK) causes defects in secondary lymphoid tissue chemokine receptor signaling and homing of peritoneal cells to the gut-associated lymphatic tissue system. J Exp Med 191:1477–1486
93. Shinkura R, Kitada K, Matsuda F, Tashiro K, Ikuta K, Suzuki M, Kogishi K, Serikawa T, Honjo T (1999) Alymphoplasia is caused by a point mutation in the mouse gene encoding Nf-kappa b-inducing kinase. Nat Genet 22:74–77
94. Li ZW, Chu W, Hu Y, Delhase M, Deerinck T, Ellisman M, Johnson R, Karin M (1999) The IKKbeta subunit of IkappaB kinase (IKK) is essential for nuclear factor kappaB activation and prevention of apoptosis. J Exp Med 189:1839–1845
95. Hu Y, Baud V, Delhase M, Zhang P, Deerinck T, Ellisman M, Johnson R, Karin M (1999) Abnormal morphogenesis but intact IKK activation in mice lacking the IKKalpha subunit of IkappaB kinase. Science 284:316–320
96. Gerondakis S, Grumont R, Gugasyan R, Wong L, Isomura I, Ho W, Banerjee A (2006) Unravelling the complexities of the NF-kappaB signalling pathway using mouse knockout and transgenic models. Oncogene 25:6781–6799
97. Tanaka M, Fuentes ME, Yamaguchi K, Durnin MH, Dalrymple SA, Hardy KL, Goeddel DV (1999) Embryonic lethality, liver degeneration, and impaired NF-kappa B activation in IKK-beta-deficient mice. Immunity 10:421–429
98. Senftleben U, Li ZW, Baud V, Karin M (2001) IKKbeta is essential for protecting T cells from TNFalpha-induced apoptosis. Immunity 14:217–230
99. Takeda K et al (1999) Limb and skin abnormalities in mice lacking IKKα. Science 284:313–316
100. Li Q, Lu Q, Hwang JY, Büscher D, Lee K-F, Izpisua-Belmonte JC, Verma IM (1999) IKK1-deficient mice exhibit abnormal development of skin and skeleton. Genes Dev 13:1322–1328
101. Solt LA, Madge LA, Orange JS, May MJ (2007) Interleukin-1-induced NF-κB activation is NEMO-dependent but does not require IKKβ. J Biol Chem 282:8724–8733

102. Schmidt-Supprian M, Bloch W, Courtois G, Addicks K, Israel A, Rajewsky K, Pasparakis M (2000) NEMO/IKK gamma-deficient mice model incontinentia pigmenti. Mol Cell 5:981–992
103. Li Q, Estepa G, Memet S, Israel A, Verma IM (2000) Complete lack of NF-κB activity in IKK1 and IKK2 double-deficient mice: additional defect in neurulation. Genes Dev 14:1729–1733
104. Hayden MS, Ghosh S (2012) NF-kappaB, the first quarter-century: remarkable progress and outstanding questions. Genes Dev 26:203–234
105. DiDonato JA, Hayakawa M, Rothwarf DM, Zandi E, Karin M (1997) A cytokine-responsive IkappaB kinase that activates the transcription factor NF-kappaB. Nature 388:548–554
106. Hacker H, Karin M (2006) Regulation and function of IKK and IKK-related kinases. Sci STKE 2006(357):re13
107. Ling L, Cao Z, Goeddel DV (1998) NF-κB-inducing kinase activates IKK-α by phosphorylation of Ser-176. Proc Natl Acad Sci 95:3792–3797
108. Regnier CH, Song HY, Gao X, Goeddel DV, Cao Z, Rothe M (1997) Identification and characterization of an IkappaB kinase. Cell 90:373–383
109. Xu G, Lo YC, Li Q, Napolitano G, Wu X, Jiang X, Dreano M, Karin M, Wu H (2011) Crystal structure of inhibitor of kappaB kinase beta. Nature 472:325–330
110. Hayden MS, Ghosh S (2004) Signaling to NF-kappaB. Genes Dev 18:2195–2224
111. Tang ED, Inohara N, Wang C-Y, Nuñez G, Guan K-L (2003) Roles for homotypic interactions and transautophosphorylation in IκB kinase (IKKβ) activation. J Biol Chem 278:38566–38570
112. Poyet J-L, Srinivasula SM, J-h L, Fernandes-Alnemri T, Yamaoka S, Tsichlis PN, Alnemri ES (2000) Activation of the IκB kinases by RIP via IKKγ/NEMO-mediated oligomerization. J Biol Chem 275:37966–37977
113. Inohara N, Koseki T, Lin J, del Peso L, Lucas PC, Chen FF, Ogura Y, Núñez G (2000) An induced proximity model for NF-κB activation in the Nod1/RICK and RIP signaling pathways. J Biol Chem 275:27823–27831
114. Wertz IE, Dixit VM (2010) Signaling to NF-kappaB: regulation by ubiquitination. Cold Spring Harb Perspect Biol 2:a003350
115. Chen J, Chen ZJ (2013) Regulation of NF-kappaB by ubiquitination. Curr Opin Immunol 25:4–12
116. Finley D, Ozkaynak E, Varshavsky A (1987) The yeast polyubiquitin gene is essential for resistance to high temperatures, starvation, and other stresses. Cell 48:1035–1046
117. Liu S, Chen ZJ (2011) Expanding role of ubiquitination in NF-[kappa]B signaling. Cell Res 21:6–21
118. Schwartz AL, Ciechanover A (2009) Targeting proteins for destruction by the ubiquitin system: implications for human pathobiology. Annu Rev Pharmacol Toxicol 49:73–96
119. Vallabhapurapu S, Karin M (2009) Regulation and function of NF-κB transcription factors in the immune system. Annu Rev Immunol 27:693–733
120. Hurley JH, Lee S, Prag G (2006) Ubiquitin-binding domains. Biochem J 399:361–372
121. Hicke L, Schubert HL, Hill CP (2005) Ubiquitin-binding domains. Nat Rev Mol Cell Biol 6:610–621
122. Ikeda F, Dikic I (2008) Atypical ubiquitin chains: new molecular signals. "Protein modifications: beyond the usual suspects" review series. EMBO Rep 9:536–542
123. Pickart CM, Fushman D (2004) Polyubiquitin chains: polymeric protein signals. Curr Opin Chem Biol 8:610–616
124. Chen ZJ, Sun LJ (2009) Nonproteolytic functions of ubiquitin in cell signaling. Mol Cell 33:275–286
125. Skaug B, Jiang X, Chen ZJ (2009) The role of ubiquitin in NF-κB regulatory pathways. Annu Rev Biochem 78:769–796
126. Reyes-Turcu FE, Wilkinson KD (2009) Polyubiquitin binding and disassembly by deubiquitinating enzymes. Chem Rev 109:1495–1508

127. Lee EG, Boone DL, Chai S, Libby SL, Chien M, Lodolce JP, Ma A (2000) Failure to regulate TNF-induced NF-κB and cell death responses in A20-deficient mice. Science 289:2350–2354
128. Compagno M et al (2009) Mutations of multiple genes cause deregulation of NF-kappaB in diffuse large B-cell lymphoma. Nature 459:717–721
129. Kato M et al (2009) Frequent inactivation of A20 in B-cell lymphomas. Nature 459:712–716
130. Novak U et al (2009) The NF-κB negative regulator TNFAIP3 (A20) is inactivated by somatic mutations and genomic deletions in marginal zone lymphomas. Blood 113:4918–4921
131. Schmitz R et al (2009) TNFAIP3 (A20) is a tumor suppressor gene in Hodgkin lymphoma and primary mediastinal B cell lymphoma. J Exp Med 206:981–989
132. Musone SL et al (2008) Multiple polymorphisms in the TNFAIP3 region are independently associated with systemic lupus erythematosus. Nat Genet 40:1062–1064
133. Sun SC (2010) CYLD: a tumor suppressor deubiquitinase regulating NF-kappaB activation and diverse biological processes. Cell Death Differ 17:25–34
134. Hershko A (1983) Ubiquitin: roles in protein modification and breakdown. Cell 34:11–12
135. Pickart CM (2001) Ubiquitin enters the new millennium. Mol Cell 8:499–504
136. Ciechanover A, Finley D, Varshavsky A (1984) Ubiquitin dependence of selective protein degradation demonstrated in the mammalian cell cycle mutant ts85. Cell 37:57–66
137. Finley D, Ciechanover A, Varshavsky A (1984) Thermolability of ubiquitin-activating enzyme from the mammalian cell cycle mutant ts85. Cell 37:43–55
138. Pomerantz JL, Baltimore D (2002) Two pathways to NF-kappaB. Mol Cell 10:693–695
139. Maniatis T (1999) A ubiquitin ligase complex essential for the NF-kappaB, Wnt/Wingless, and Hedgehog signaling pathways. Genes Dev 13:505–510
140. Lin L, Ghosh S (1996) A glycine-rich region in NF-kappaB p105 functions as a processing signal for the generation of the p50 subunit. Mol Cell Biol 16:2248–2254
141. Piwko W, Jentsch S (2006) Proteasome-mediated protein processing by bidirectional degradation initiated from an internal site. Nat Struct Mol Biol 13:691–697
142. McKenzie FR, Connelly MA, Balzarano D, Muller JR, Geleziunas R, Marcu KB (2000) Functional isoforms of IkappaB kinase alpha (IKKalpha) lacking leucine zipper and helix-loop-helix domains reveal that IKKalpha and IKKbeta have different activation requirements. Mol Cell Biol 20:2635–2649
143. Agou F, Ye F, Goffinont S, Courtois G, Yamaoka S, Israël A, Véron M (2002) NEMO trimerizes through its coiled-coil C-terminal domain. J Biol Chem 277:17464–17475
144. Tegethoff S, Behlke J, Scheidereit C (2003) Tetrameric oligomerization of IkappaB kinase gamma (IKKgamma) is obligatory for IKK complex activity and NF-kappaB activation. Mol Cell Biol 23:2029–2041
145. Drew D, Shimada E, Huynh K, Bergqvist S, Talwar R, Karin M, Ghosh G (2007) Inhibitor kappaB kinase beta binding by inhibitor kappaB kinase gamma. Biochemistry 46:12482–12490
146. Agou F, Traincard F, Vinolo E, Courtois G, Yamaoka S, Israël A, Véron M (2004) The trimerization domain of NEMO is composed of the interacting C-terminal CC2 and LZ coiled-coil subdomains. J Biol Chem 279:27861–27869
147. Poyet J-L, Srinivasula SM, Alnemri ES (2001) vCLAP, a caspase-recruitment domain-containing protein of equine Herpesvirus-2, persistently activates the IκB kinases through oligomerization of IKKγ. J Biol Chem 276:3183–3187
148. Huang GJ, Zhang ZQ, Jin DY (2002) Stimulation of IKK-gamma oligomerization by the human T-cell leukemia virus oncoprotein Tax. FEBS Lett 531:494–498
149. Hofmann RM, Pickart CM (2001) In vitro assembly and recognition of Lys-63 polyubiquitin chains. J Biol Chem 276:27936–27943
150. Xu P et al (2009) Quantitative proteomics reveals the function of unconventional ubiquitin chains in proteasomal degradation. Cell 137:133–145
151. Chen ZJ (2012) Ubiquitination in signaling to and activation of IKK. Immunol Rev 246:95–106
152. Deng L, Wang C, Spencer E, Yang L, Braun A, You J, Slaughter C, Pickart C, Chen ZJ (2000) Activation of the IkappaB kinase complex by TRAF6 requires a dimeric ubiquitin-conjugating enzyme complex and a unique polyubiquitin chain. Cell 103:351–361

153. Hofmann RM, Pickart CM (1999) Noncanonical MMS2-encoded ubiquitin-conjugating enzyme functions in assembly of novel polyubiquitin chains for DNA repair. Cell 96:645–653
154. Lomaga MA et al (1999) TRAF6 deficiency results in osteopetrosis and defective interleukin-1, CD40, and LPS signaling. Genes Dev 13:1015–1024
155. Naito A et al (1999) Severe osteopetrosis, defective interleukin-1 signalling and lymph node organogenesis in TRAF6-deficient mice. Genes Cells 4:353–362
156. Ea CK, Deng L, Xia ZP, Pineda G, Chen ZJ (2006) Activation of IKK by TNFalpha requires site-specific ubiquitination of RIP1 and polyubiquitin binding by NEMO. Mol Cell 22:245–257
157. Wu CJ, Conze DB, Li T, Srinivasula SM, Ashwell JD (2006) Sensing of Lys 63-linked polyubiquitination by NEMO is a key event in NF-kappaB activation [corrected]. Nat Cell Biol 8:398–406
158. Zhou H, Wertz I, O'Rourke K, Ultsch M, Seshagiri S, Eby M, Xiao W, Dixit VM (2004) Bcl10 activates the NF-kappaB pathway through ubiquitination of NEMO. Nature 427:167–171
159. Ni CY et al (2008) Cutting edge: K63-linked polyubiquitination of NEMO modulates TLR signaling and inflammation in vivo. J Immunol 180:7107–7111
160. Wang C, Deng L, Hong M, Akkaraju GR, Inoue J, Chen ZJ (2001) TAK1 is a ubiquitin-dependent kinase of MKK and IKK. Nature 412:346–351
161. Xia ZP, Sun L, Chen X, Pineda G, Jiang X, Adhikari A, Zeng W, Chen ZJ (2009) Direct activation of protein kinases by unanchored polyubiquitin chains. Nature 461:114–119
162. Shim J-H et al (2005) TAK1, but not TAB1 or TAB2, plays an essential role in multiple signaling pathways in vivo. Genes Dev 19:2668–2681
163. Chen ZJ, Bhoj V, Seth RB (2006) Ubiquitin, TAK1 and IKK: is there a connection? Cell Death Differ 13:687–692
164. Oeckinghaus A, Wegener E, Welteke V, Ferch U, Arslan SC, Ruland J, Scheidereit C, Krappmann D (2007) Malt1 ubiquitination triggers NF-kappaB signaling upon T-cell activation. EMBO J 26:4634–4645
165. Windheim M, Stafford M, Peggie M, Cohen P (2008) Interleukin-1 (IL-1) induces the Lys63-linked polyubiquitination of IL-1 receptor-associated kinase 1 to facilitate NEMO binding and the activation of IkappaBalpha kinase. Mol Cell Biol 28:1783–1791
166. Grivennikov SI, Karin M (2010) Inflammation and oncogenesis: a vicious connection. Curr Opin Genet Dev 20:65–71
167. Lawrence T (2009) The nuclear factor NF-kappaB pathway in inflammation. Cold Spring Harb Perspect Biol 1:a001651
168. Staudt LM (2010) Oncogenic activation of NF-kappaB. Cold Spring Harb Perspect Biol 2:a000109
169. Ruland J (2011) Return to homeostasis: downregulation of NF-[kappa]B responses. Nat Immunol 12:709–714
170. Arenzana-Seisdedos F, Thompson J, Rodriguez MS, Bachelerie F, Thomas D, Hay RT (1995) Inducible nuclear expression of newly synthesized I kappa B alpha negatively regulates DNA-binding and transcriptional activities of NF-kappa B. Mol Cell Biol 15:2689–2696
171. Peng B et al (2010) Defective feedback regulation of NF-kappaB underlies Sjogren's syndrome in mice with mutated kappaB enhancers of the IkappaBalpha promoter. Proc Natl Acad Sci U S A 107:15193–15198
172. Ghosh S, Hayden MS (2012) Celebrating 25 years of NF-kappaB research. Immunol Rev 246:5–13
173. Rao P et al (2010) IkappaBbeta acts to inhibit and activate gene expression during the inflammatory response. Nature 466:1115–1119
174. Scheibel M, Klein B, Merkle H, Schulz M, Fritsch R, Greten FR, Arkan MC, Schneider G, Schmid RM (2010) IkappaBbeta is an essential co-activator for LPS-induced IL-1beta transcription in vivo. J Exp Med 207:2621–2630
175. Phillips RJ, Ghosh S (1997) Regulation of IkappaB beta in WEHI 231 mature B cells. Mol Cell Biol 17:4390–4396

176. Weil R, Laurent-Winter C, Israël A (1997) Regulation of IκBβ degradation. Similarities to and differences from IκBα. J Biol Chem 272:9942–9949
177. Bergqvist S, Alverdi V, Mengel B, Hoffmann A, Ghosh G, Komives EA (2009) Kinetic enhancement of NF-kappaBxDNA dissociation by IkappaBalpha. Proc Natl Acad Sci U S A 106:19328–19333
178. Tran K, Merika M, Thanos D (1997) Distinct functional properties of IkappaB alpha and IkappaB beta. Mol Cell Biol 17:5386–5399
179. Malek S, Huang D-B, Huxford T, Ghosh S, Ghosh G (2003) X-ray crystal structure of an IκBβ·NF-κB p65 homodimer complex. J Biol Chem 278:23094–23100
180. Mahoney DJ et al (2008) Both cIAP1 and cIAP2 regulate TNFalpha-mediated NF-kappaB activation. Proc Natl Acad Sci U S A 105:11778–11783
181. Kawai T, Akira S (2010) The role of pattern-recognition receptors in innate immunity: update on Toll-like receptors. Nat Immunol 11:373–384
182. Rawlings DJ, Sommer K, Moreno-Garcia ME (2006) The CARMA1 signalosome links the signalling machinery of adaptive and innate immunity in lymphocytes. Nat Rev Immunol 6:799–812
183. Wertz IE et al (2004) De-ubiquitination and ubiquitin ligase domains of A20 downregulate NF-kappaB signalling. Nature 430:694–699
184. Newton K et al (2008) Ubiquitin chain editing revealed by polyubiquitin linkage-specific antibodies. Cell 134:668–678
185. Heyninck K, Beyaert R (2005) A20 inhibits NF-kappaB activation by dual ubiquitin-editing functions. Trends Biochem Sci 30:1–4
186. Boone DL et al (2004) The ubiquitin-modifying enzyme A20 is required for termination of Toll-like receptor responses. Nat Immunol 5:1052–1060
187. Vereecke L, Beyaert R, van Loo G (2009) The ubiquitin-editing enzyme A20 (TNFAIP3) is a central regulator of immunopathology. Trends Immunol 30:383–391
188. Opipari AW Jr, Boguski MS, Dixit VM (1990) The A20 cDNA induced by tumor necrosis factor alpha encodes a novel type of zinc finger protein. J Biol Chem 265:14705–14708
189. Dixit VM, Green S, Sarma V, Holzman LB, Wolf FW, O'Rourke K, Ward PA, Prochownik EV, Marks RM (1990) Tumor necrosis factor-alpha induction of novel gene products in human endothelial cells including a macrophage-specific chemotaxin. J Biol Chem 265:2973–2978
190. Evans PC, Ovaa H, Hamon M, Kilshaw PJ, Hamm S, Bauer S, Ploegh HL, Smith TS (2004) Zinc-finger protein A20, a regulator of inflammation and cell survival, has de-ubiquitinating activity. Biochem J 378:727–734
191. Mauro C, Pacifico F, Lavorgna A, Mellone S, Iannetti A, Acquaviva R, Formisano S, Vito P, Leonardi A (2006) ABIN-1 binds to NEMO/IKKgamma and co-operates with A20 in inhibiting NF-kappaB. J Biol Chem 281:18482–18488
192. Hitotsumatsu O et al (2008) The ubiquitin-editing enzyme A20 restricts nucleotide-binding oligomerization domain containing 2-triggered signals. Immunity 28:381–390
193. Duwel M et al (2009) A20 negatively regulates T cell receptor signaling to NF-kappaB by cleaving Malt1 ubiquitin chains. J Immunol 182:7718–7728
194. Song HY, Rothe M, Goeddel DV (1996) The tumor necrosis factor-inducible zinc finger protein A20 interacts with TRAF1/TRAF2 and inhibits NF-kappaB activation. Proc Natl Acad Sci U S A 93:6721–6725
195. Komander D, Barford D (2008) Structure of the A20 OTU domain and mechanistic insights into deubiquitination. Biochem J 409:77–85
196. Lin SC, Chung JY, Lamothe B, Rajashankar K, Lu M, Lo YC, Lam AY, Darnay BG, Wu H (2008) Molecular basis for the unique deubiquitinating activity of the NF-kappaB inhibitor A20. J Mol Biol 376:526–540
197. Enesa K, Zakkar M, Chaudhury H, Luong LA, Rawlinson L, Mason JC, Haskard DO, Dean JLE, Evans PC (2008) NF-κB suppression by the deubiquitinating enzyme cezanne: a novel negative feedback loop in pro-inflammatory signaling. J Biol Chem 283:7036–7045

198. Bremm A, Freund SM, Komander D (2010) Lys11-linked ubiquitin chains adopt compact conformations and are preferentially hydrolyzed by the deubiquitinase Cezanne. Nat Struct Mol Biol 17:939–947
199. Trompouki E, Hatzivassiliou E, Tsichritzis T, Farmer H, Ashworth A, Mosialos G (2003) CYLD is a deubiquitinating enzyme that negatively regulates NF-kappaB activation by TNFR family members. Nature 424:793–796
200. Kovalenko A, Chable-Bessia C, Cantarella G, Israel A, Wallach D, Courtois G (2003) The tumour suppressor CYLD negatively regulates NF-kappaB signalling by deubiquitination. Nature 424:801–805
201. Wright A, Reiley WW, Chang M, Jin W, Lee AJ, Zhang M, Sun SC (2007) Regulation of early wave of germ cell apoptosis and spermatogenesis by deubiquitinating enzyme CYLD. Dev Cell 13:705–716
202. Zhang J, Stirling B, Temmerman ST, Ma CA, Fuss IJ, Derry JM, Jain A (2006) Impaired regulation of NF-kappaB and increased susceptibility to colitis-associated tumorigenesis in CYLD-deficient mice. J Clin Invest 116:3042–3049
203. Reiley WW et al (2007) Deubiquitinating enzyme CYLD negatively regulates the ubiquitin-dependent kinase Tak1 and prevents abnormal T cell responses. J Exp Med 204:1475–1485
204. Komander D, Reyes-Turcu F, Licchesi JD, Odenwaelder P, Wilkinson KD, Barford D (2009) Molecular discrimination of structurally equivalent Lys 63-linked and linear polyubiquitin chains. EMBO Rep 10:466–473
205. Komander D, Lord CJ, Scheel H, Swift S, Hofmann K, Ashworth A, Barford D (2008) The structure of the CYLD USP domain explains its specificity for Lys63-linked polyubiquitin and reveals a B box module. Mol Cell 29:451–464
206. Reiley WW, Zhang M, Jin W, Losiewicz M, Donohue KB, Norbury CC, Sun SC (2006) Regulation of T cell development by the deubiquitinating enzyme CYLD. Nat Immunol 7:411–417
207. Sun SC (2008) Deubiquitylation and regulation of the immune response. Nat Rev Immunol 8:501–511
208. Massoumi R, Chmielarska K, Hennecke K, Pfeifer A, Fassler R (2006) Cyld inhibits tumor cell proliferation by blocking Bcl-3-dependent NF-kappaB signaling. Cell 125:665–677
209. Reiley W, Zhang M, Wu X, Granger E, Sun SC (2005) Regulation of the deubiquitinating enzyme CYLD by IkappaB kinase gamma-dependent phosphorylation. Mol Cell Biol 25:3886–3895
210. Jin W et al (2008) Deubiquitinating enzyme CYLD negatively regulates RANK signaling and osteoclastogenesis in mice. J Clin Invest 118:1858–1866
211. Jin W, Reiley WR, Lee AJ, Wright A, Wu X, Zhang M, Sun SC (2007) Deubiquitinating enzyme CYLD regulates the peripheral development and naive phenotype maintenance of B cells. J Biol Chem 282:15884–15893
212. Lee AJ, Zhou X, Chang M, Hunzeker J, Bonneau RH, Zhou D, Sun SC (2010) Regulation of natural killer T-cell development by deubiquitinase CYLD. EMBO J 29:1600–1612
213. Hovelmeyer N et al (2007) Regulation of B cell homeostasis and activation by the tumor suppressor gene CYLD. J Exp Med 204:2615–2627
214. Makris C, Godfrey VL, Krahn-Senftleben G, Takahashi T, Roberts JL, Schwarz T, Feng L, Johnson RS, Karin M (2000) Female mice heterozygous for IKK gamma/NEMO deficiencies develop a dermatopathy similar to the human X-linked disorder incontinentia pigmenti. Mol Cell 5:969–979
215. Li H, Kobayashi M, Blonska M, You Y, Lin X (2006) Ubiquitination of RIP is required for tumor necrosis factor α-induced NF-κB activation. J Biol Chem 281:13636–13643
216. Xu M, Skaug B, Zeng W, Chen ZJ (2009) A ubiquitin replacement strategy in human cells reveals distinct mechanisms of IKK activation by TNFalpha and IL-1beta. Mol Cell 36:302–314
217. Fukushima T, S-i M, Kress CL, Bruey JM, Krajewska M, Lefebvre S, Zapata JM, Ze R, Reed JC (2007) Ubiquitin-conjugating enzyme Ubc13 is a critical component of TNF receptor-associated factor (TRAF)-mediated inflammatory responses. Proc Natl Acad Sci 104:6371–6376

218. Yamamoto M et al (2006) Key function for the Ubc13 E2 ubiquitin-conjugating enzyme in immune receptor signaling. Nat Immunol 7:962–970
219. Yeh WC et al (1997) Early lethality, functional NF-kappaB activation, and increased sensitivity to TNF-induced cell death in TRAF2-deficient mice. Immunity 7:715–725
220. Nakano H et al (1999) Targeted disruption of Traf5 gene causes defects in CD40- and CD27-mediated lymphocyte activation. Proc Natl Acad Sci U S A 96:9803–9808
221. Tada K et al (2001) Critical roles of TRAF2 and TRAF5 in tumor necrosis factor-induced NF-kappa B activation and protection from cell death. J Biol Chem 276:36530–36534
222. Kobayashi N, Kadono Y, Naito A, Matsumoto K, Yamamoto T, Tanaka S, Inoue J (2001) Segregation of TRAF6-mediated signaling pathways clarifies its role in osteoclastogenesis. EMBO J 20:1271–1280
223. Conze DB, Albert L, Ferrick DA, Goeddel DV, Yeh WC, Mak T, Ashwell JD (2005) Posttranscriptional downregulation of c-IAP2 by the ubiquitin protein ligase c-IAP1 in vivo. Mol Cell Biol 25:3348–3356
224. Conte D, Holcik M, Lefebvre CA, Lacasse E, Picketts DJ, Wright KE, Korneluk RG (2006) Inhibitor of apoptosis protein cIAP2 is essential for lipopolysaccharide-induced macrophage survival. Mol Cell Biol 26:699–708
225. Varfolomeev E, Vucic D (2008) (Un)expected roles of c-IAPs in apoptotic and NFkappaB signaling pathways. Cell Cycle 7:1511–1521
226. Kelliher MA, Grimm S, Ishida Y, Kuo F, Stanger BZ, Leder P (1998) The death domain kinase RIP mediates the TNF-induced NF-kappaB signal. Immunity 8:297–303
227. Gerlach B et al (2011) Linear ubiquitination prevents inflammation and regulates immune signalling. Nature 471:591–596
228. Dynek JN et al (2010) c-IAP1 and UbcH5 promote K11-linked polyubiquitination of RIP1 in TNF signalling. EMBO J 29:4198–4209
229. Haas TL et al (2009) Recruitment of the linear ubiquitin chain assembly complex stabilizes the TNF-R1 signaling complex and is required for TNF-mediated gene induction. Mol Cell 36:831–844
230. Mollah S, Wertz IE, Phung Q, Arnott D, Dixit VM, Lill JR (2007) Targeted mass spectrometric strategy for global mapping of ubiquitination on proteins. Rapid Commun Mass Spectrom 21:3357–3364
231. Kim W et al (2011) Systematic and quantitative assessment of the ubiquitin-modified proteome. Mol Cell 44:325–340
232. Wong WW, Gentle IE, Nachbur U, Anderton H, Vaux DL, Silke J (2010) RIPK1 is not essential for TNFR1-induced activation of NF-kappaB. Cell Death Differ 17:482–487
233. Zhang H, Zhou X, McQuade T, Li J, Chan FK, Zhang J (2011) Functional complementation between FADD and RIP1 in embryos and lymphocytes. Nature 471:373–376
234. Ofengeim D, Yuan J (2013) Regulation of RIP1 kinase signalling at the crossroads of inflammation and cell death. Nat Rev Mol Cell Biol 14:727–736

Chapter 2
Regulation of NF-κB Signaling by Regulatory Ubiquitination: Specific Pathways and A20

2.1 Regulatory Ubiquitination and NF-κB

Given the importance of IKK in phosphorylating IκBs, it is imperative to understand how IKK is activated. There is no doubt that degradative ubiquitination of the IκBs (and other molecules) plays a profound role in NF-κB activation. One of the first clues that unconventional ubiquitin chains may also have a role to play in activating IKK came when it was found that IKK was activated by an E2 of the Ubc4/5 family [1], and that this activation was independent of proteasomal activity and the K48 residue of ubiquitin [2]. While this suggested the involvement of a proteasome-independent mechanism, the cellular relevance of this finding was unclear as the specific E3 ligase had not yet been identified. It was later determined that the ubiquitin ligase was TRAF6 [3], a protein that also functions as an adaptor molecule in the IL-1 and Toll-like receptor pathways [4, 5].

Subsequently though (and as detailed below), it has been shown that autoubiquitination of TRAF6 is not required for activation of NF-κB and MAPK pathways [6, 7]. While it is theoretically possible that the ubiquitin chains may provide a docking site for formation of the TAK–TAB signaling complex, it was shown using a TRAF6-RING-mutant that ubiquitination was dispensable for the recruitment of the TAK–TAB complex. Thus, the activity of TRAF6 as an adaptor protein (independent of ubiquitination) is enough to recruit the complex. The authors concluded that while TRAF6 ubiquitination may be a marker for activation, it is unlikely to play an essential role for downstream activation. Recent findings, mostly in vitro work, have advanced the proposition that K63-linked polyubiquitin chains may have a role to play in activating IKK in multiple pathways that activate NF-κB [8, 9].

The physiological targets of K63-linked polyubiquitination have not yet been identified [10]. The physiological context of regulatory ubiquitination can be clarified by generation of RING-mutant or deubiquitinating-mutant knock-in mice [9], something that has not been done for any of the enzymes involved in these pathways. In spite of the overwhelming biochemical evidence, genetic evidence that

A. De, *Ubiquitin Chains: Degradation and Beyond*, Springer Theses,
DOI 10.1007/978-3-319-14965-3_2

points to the specific role of regulatory ubiquitination is lacking. This is partly because all the experiments have been done either by knocking down or overexpressing the relevant proteins. This has been ineffective as it could lead to oligomerization of TRAFs, and/or, induce promiscuity in these enzymes.

An example of this limitation is highlighted by knock-in studies in NEMO. Polyubiquitination of K399 in NEMO was shown to be essential for NF-κB activation by an experiment based on overexpression with BCL10. The K399R mutant was not ubiquitinated and NF-κB was only partially activated in NEMO-deficient cell lines complemented with K399R [11]. However a murine knock-in model carrying this mutation was found to be largely normal with respect to NF-κB and MAPK activation [12]. While outlining the various signaling pathways below, I will be highlighting such discrepancies in order to have an accurate understanding of the present status of the field.

2.2 IL-1R/TLR Signaling Pathways

TLR (Toll-like receptors) are PRRs that recognize various PAMPs that are present in different microbes (bacteria and viruses) including bacterial LPS and viral nucleic acids. Interleukin-1 (IL-1) is a family of cytokines secreted by cells during the inflammatory process and this serves as a defensive response during pathogenic invasion [13]. Upon ligand binding, both pathways activate subsets of proinflammatory genes. Both TLR and interleukin-1 receptor (IL1R) are transmembrane proteins that have a common cytosolic domain called Toll-IL-1R (TIR) domain. Following stimulation, this domain helps recruit TIR-domain adaptor proteins that can bind to the TIR domain of TLR and IL1R. The most common example of such an adaptor is the myeloid differentiation primary gene 88 (MyD88). MyD88 then recruits two kinases to the IL1R, IRAK4, and IRAK1. IRAK1 binds TRAF6, which together with Ubc13/Uev1A catalyzes the formation of K63-linked polyubiquitin chains. Both IRAK1 and TRAF6 have been reported to be modified by K63-linked chains. The polyubiquitin chains on TRAF6 recruits the TAK1–TAB2–TAB3 complex (a protein complex of TGF-β activated kinase 1 with adaptor proteins TAB1 and TAB2) [14, 15]. The complex binds the ubiquitin chains with the UBD of the TAB proteins. This recruitment of the TAK–TAB complex leads to the activation of TAK1 [16]. Subsequently, TAK1 phosphorylates other downstream kinases like IKK (leading to NF-κB activation) or MKKs (leading to the JNK and p38 activation).

Similar pathways are also activated in binding of TLR ligands. As an example, let us consider the binding of LPS to TLR4 as the mechanistic details have been confirmed by multiple studies. LPS binding activates two branches downstream of TLR4, both of which lead to activation of TAK1, which subsequently phosphorylates and activates IKK. In one branch, receptor binding leads to the recruitment of TRAM and TRIF proteins to TLR4, which then recruits both TRAF6 and RIP1 to the signaling complex. In the other branch, receptor binding leads to Myd88 recruitment which then promotes the assembly of IRAK proteins, TRAF6 and

TAK–TAB complexes as described above. It has been proposed almost entirely on the basis of in vitro evidence that K63-linked polyubiquitination of TRAF6 and RIP1 play an essential role in downstream activation of NF-κB [9].

However the genetic evidence is unsatisfactory as TAB proteins have not been found to be physiologically essential, as would be critical if the activation of TAK were indeed to depend on TAB binding to ubiquitin chains [17]. Additionally, it has been showed that Ubc13 knockouts have impaired activation of JNK and p38 kinase, but normal activation of IKK [18]. Thus NF-κB signaling is normal in UBC13-deficient B cells, bone marrow-derived macrophages, or MEFs when stimulated with IL1-β or multiple TLR ligands including LPS, bacterial lipopeptide, and CpG DNA. Subsequently though, in another study, UBC13 was shown to be important in IKK activation [19–22]. The reason for this inconsistency has been speculated to be incomplete deletion of Ubc13 by Cre recombinase in the first study. Indeed, titrating the Cre retroviruses and studying the effects of variable Ubc13 depletion, the authors concluded [21] that a minor amount of residual Ubc13 was enough to activate IKK. This however raises doubts on whether the K63 chains indeed play an actual role in IKK activation or if the ubiquitin chains result from aggregation of residual Ubc13 with TRAF6 during NF-κB activation. In this regard, multiple proteins that associate with TRAF6 have been shown to be ubiquitinated by K63 chains [15, 23–26]. It is not clear if their ubiquitination is actually intrinsic for the act of signaling.

Along with a lack of genetic evidence, certain biochemical studies have also suggested that ubiquitination of TRAF6 may not be essential. For example, a study found that a ligase-mutant of TRAF6 (i.e., TRAF6 with RING domain deleted) [6] did not block IKK activation by IL-1β. However, this mutant failed to activate JNK and TAK1. In addition to demonstrating that the ubiquitin ligase activity of TRAF6 is not required for IKK activation, this study also suggested that TAK1 may not be the kinase for IKK, and that IKK could activate itself by autophosphorylation. A second independent study confirmed that NF-κB (and MAPK) pathways could be activated by IL-1 and that this does not require TRAF6 autoubiquitination. Thus, complementation of TRAF6-knockout MEFs with a TRAF6 mutant that lacks lysines, and hence cannot be ubiquitinated, shows normal IKK activation. This mutant TRAF6 is physiologically active and rescues osteoclastogenesis when retrovirally transduced in TRAF6-knockout bone marrow macrophages. This may be because the lysine-deficient TRAF6 mutant could serve as an adaptor protein and interact with the TAK1–TAB1–TAB2 complex [7]. Subsequently, other conflicting studies have demonstrated the importance of the RING domain of TRAF6 in IKK activation via the IL-1-mediated pathway [21, 23, 27–29].

Additional evidence for the proposed mechanism of IKK activation was obtained by an elegant system which replaced endogenous ubiquitin with K63R mutant ubiquitin following tetracycline treatment. The authors show that in cells having the K63R ubiquitin, IKK cannot be activated by IL-1β. This demonstrates the importance of K63-linked polyubiquitin chains in mediating IKK activation. Intriguingly though, the same study also shows that IKK can still be activated normally in response to TNFα even when only K63R ubiquitin is present, suggesting that IKK may be activated by TNFα by a K63-polyubiquitin-independent mechanism [19].

Recently it was demonstrated that free, unconjugated K63-polyuniquitin chains (not bound to any substrate) could also activate TAK1 [30]. The authors synthesized the K63-linked polyubiquitin chains by treatment with TRAF6 and Ubc13/Uev1A. As expected, the UBD domain of TAB2 is associated with these chains, bringing the TAK1 in close proximity. Subsequently, TAK1 was activated by phosphorylation at Thr-187. Activated TAK1 then phosphorylates IKKβ, leading to activation of IKK complex in a NEMO-dependent manner. It is not clear how the TAK1 would distinguish between polyubiquitin chains freely floating within the cell and polyubiquitin chains that are attached to TRAF6 (or for that matter, any other protein). This concern for specificity seems to be a recurring issue in the case of K63-linked polyubiquitin chain-mediated activation of IKK.

NF-κB activation is reversible; and the case for positive regulation by K63-linked regulatory ubiquitin chains is supported by the activity of deubiquitinases like CYLD and A20 that disassembles these chains, thereby negatively regulating NF-κB. There is much biochemical evidence, often based on the overexpression of the DUB that shows that deubiquitinases may block the activation of IKK by cleaving K63 chains. An example is the familial cylindromatosis tumor suppressor (CYLD), which specifically removes K63 chains from TRAF6 [26, 31–33]. Another example of a deubiquitinase is A20, which prevents NF-κB activation mediated by IL-1β or LPS by preventing the interaction between TRAF6 and Ubc13 [34–36]. While knockouts of A20 and CYLD show dysregulated NF-κB activation, it has not been demonstrated by making knock-ins that their deubiquitinase function is actually responsible for NF-κB regulation [37–39].

Figure 2.1 shows how regulatory ubiquitination could potentially regulate the NF-κB pathway downstream of IL-1R/TLR.

2.3 TNF Receptor (TNFR) Signaling Pathway

The pathway is demonstrated in Fig. 2.2. Discovered in 1975, tumor necrosis factor-α (TNF-α) was so named as it caused necrosis of murine tumors [40]. TNF-α also has potentially protective roles as a proinflammatory cytokine that stimulates multiple responses and activates NF-κB signaling. Dysregulation of these responses results in a whole host of inflammatory and autoimmune diseases [41].

TNF-α binds to two receptors that are responsible for mediating its functions. The receptors are TNFR1 and TNFR2. While TNFR1 is expressed ubiquitously in almost all mammalian cells, TNFR2 is expressed mostly in lymphocytes [42]. For the purposes of this discussion, we will only focus on signals emanating from the TNFR1 pathway. TNF-α exists in two forms: a membrane-bound form and a soluble form. The membrane-bound form is converted to the soluble form by TNFα converting enzyme (TACE). Both forms are capable of binding to the receptor. Binding of TNF-α causes trimerization of TNFR1 and consequent recruitment of TNFR1-associated death domain protein (TRADD). Remarkably, TRADD can assemble two complexes that mediate potentially opposing functions downstream.

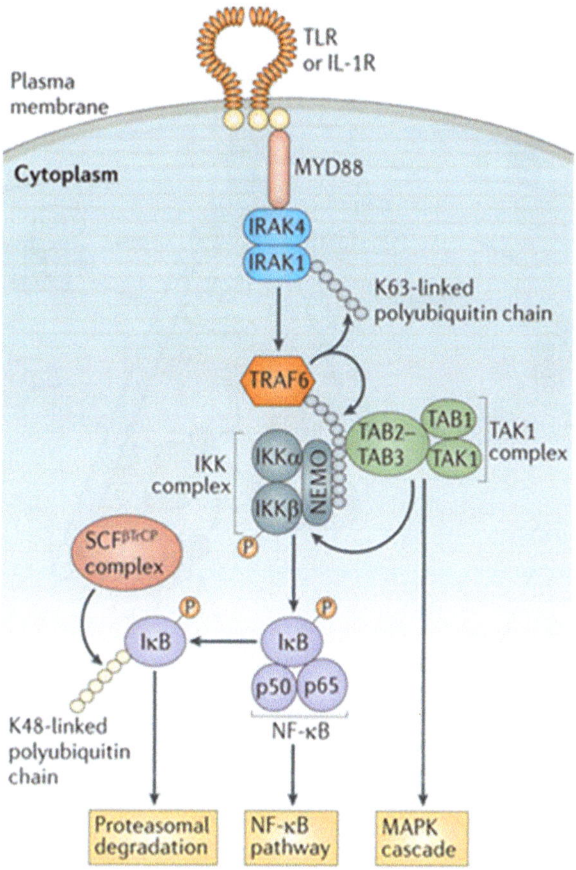

Fig. 2.1 Ubiquitin-mediated regulation in IL-1R/TLR pathways (Reprinted by permission from Macmillan Publishers Ltd: [Nature Immunology] (Xiaomo Jiang and Zhijian J. Chen (2011) The role of ubiquitylation in immune defence and pathogen evasion. Nat Immunol 12: 35-48), copyright (2011))

Complex1 mediates the activation of NF-κB, which has an antiapoptotic and proinflammatory role, while Complex 2 initiates apoptosis [43]. However, TNF-α normally does not cause apoptosis as NF-κB is rapidly activated with the consequent production of multiple antiapoptotic proteins like cIAPs and c-FLIP. Promotion of the degradation of cIAPs leads to apoptosis [44, 45].

Complex I is composed of TRADD, RIP1 (or receptor-interacting protein kinase 1), and multiple RING domain containing E3 ligases like TRAF2, TRAF5, cIAP1, and cIAP2. cIAP1 and cIAP2 catalyze K63 polyubiquitination on RIP1 [19, 46, 47]. However, cIAPs are promiscuous ubiquitin ligases and ubiquitin chains added by cIAPs are not confined to K63 alone [46, 47]. Consequently, ubiquitin chains with other linkages have been found on RIP1 (detected using antibodies specific for K48 and K63 chains) [48], and this could potentially affect the specificity of the downstream pathway.

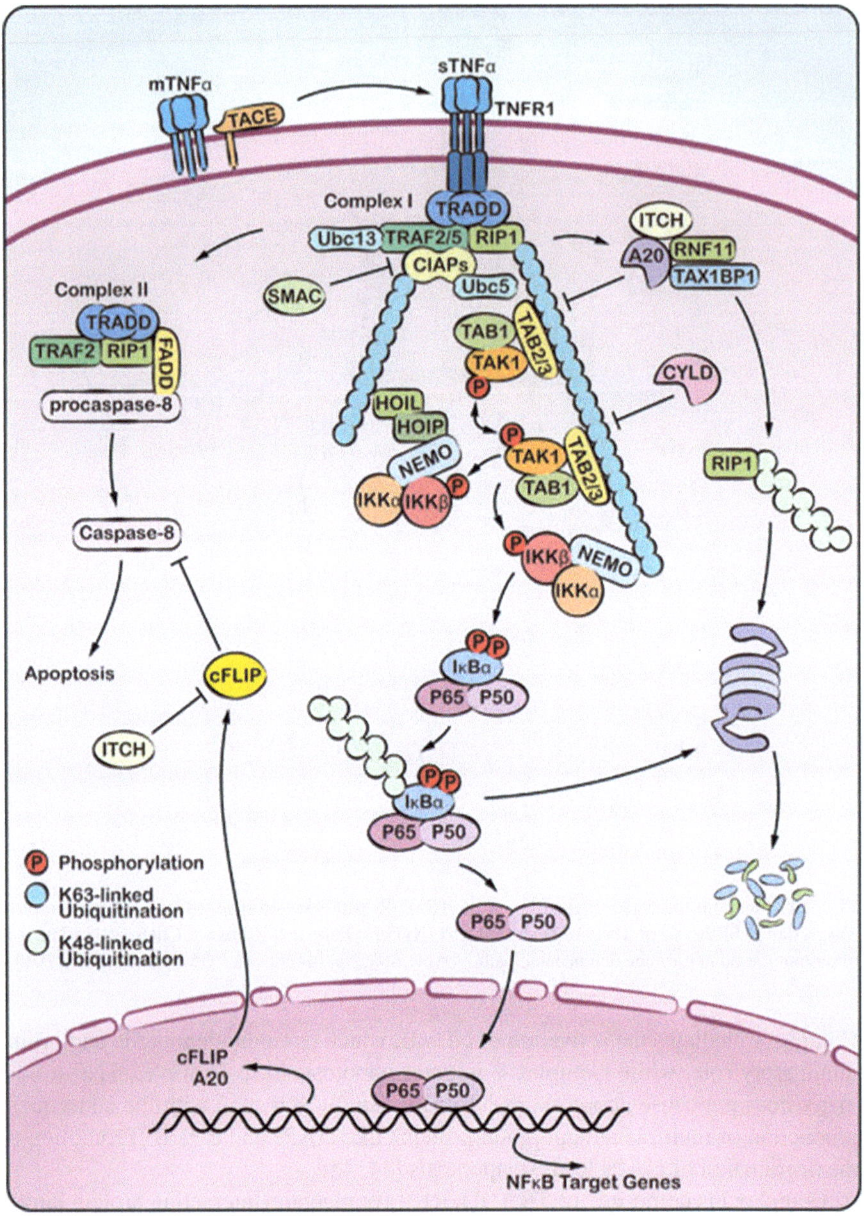

Fig. 2.2 Ubiquitin-mediated activation of TAK1 and IKK in TNFα-mediated pathway (Reprinted by permission from Macmillan Publishers Ltd: [Cell Research] (Liu S, Chen ZJ (2011) Expanding role of ubiquitination in NF-[kappa]B signaling. Cell Res 21: 6-21), copyright (2011))

RIP1 has indeed been shown to be essential for TNF-mediated NF-κB activation and RIP1 knockouts die early (1–3 days) from extensive apoptosis of lymphoid and adipose tissue because of their inability to access the NF-κB pathway [49]. While these animals do survive longer than mice deficient in IKKβ or p65, this difference may be because of the involvement of RIP1 in NF-κB independent death-pathways like necroptosis [50, 51]. However, there are conflicting studies that have reported that RIP1 may not be universally essential in all cases of TNF-induced activation of NF-κB [51, 52], and that NF-κB can still be activated in the absence of RIP1. It is possible, though, that this residual NF-κB activity is a result of the activation of the non-canonical pathway [53].

Experiments in cell lines showed RIP1 is associated with NEMO and this association is essential for TNFα-induced IKK activation [49, 51, 54, 55]. RIP1 is a kinase and given the phenotype of RIP1 knockouts, it would be a satisfactory candidate for a kinase that could activate IKK. However, the kinase activity has been shown to be dispensable for NF-κB activation [49, 55, 56]. This provided an impetus to uncover the mechanism by which RIP1 mediates the TNFα pathway.

Since RIP1 mediates the TNF pathway in a kinase-independent manner, it is tempting to speculate that it serves as an essential adaptor molecule. NEMO can bind directly to both positive and negative regulators of the NF-κB pathway. Indeed, it has been known for some time now that RIP1 can recruit IKK to the TNFR signaling complex by directly binding with NEMO [57]. NEMO can also bind with A20, which is known to negatively regulate the TNF signaling pathway [57]. There are multiple reports that the function of RIP may be to assemble a scaffold for recruitment and activation of IKK [54, 55, 58].

Ubiquitin chains may also serve as a scaffold for recruitment of receptor complexes. RIP1 ubiquitination has been proposed to be important for downstream activation of TAK1 and IKK. The TAK–TAB complex and IKK complex is recruited to Complex1 by the binding of TAB2 and NEMO to the polyubiquitin chains respectively [59]. It has been shown that RIP1 gets ubiquitinated at K377 and the K377R-RIP1 mutant fails to activate NF-κB [59]. However, the K377R-NEMO also does not get recruited to Complex1 upon receptor stimulation. Recruitment to the receptor is a prerequisite for participation in signaling. Hence, while it is clear that the K377R mutant is functionally compromised, this may not be because it cannot be ubiquitinated. It could also be because the mutant cannot serve as an effective adaptor protein. RIP1 is indeed decorated by K63-linked polyubiquitin chains upon stimulation with TNFα, and this has been directly [60] verified by using a K63-specific antibody. However, it is not known if this ubiquitination is essential for the NF-κB pathway.

There are other examples in the pathway where it is unclear as to whether the adaptor function or regulatory ubiquitination is actually required for downstream singling. Both the TNFα- and IL-1-mediated pathways seem to activate TAK1 and IKK. They also both use TRAF family adaptor proteins (TRAF2 and 5 for TNFα-mediated pathway and TRAF6 for IL-1β-mediated pathway). Given that TRAF2 and TRAF6 are structurally similar, it might be speculated that the TNF pathway could also use TRAF2 or/and TRAF5 to catalyze the ubiquitin chains, just like TRAF6 in the IL-1β pathway. Indeed TRAF2 is essential in the context of TNF-α-stimulated NF-κB activation [61, 62]. Absence of TRAF2 decreases RIP1 ubiquitination. This was seen using TRAF2 RNAi [63] and by gene ablation [56]. Genetic evidence showed that TRAF2 and TRAF5 double-knockouts have impaired TNFα signaling (but not IL-1β signaling). Additionally, the MEFs of the double-knockouts are more susceptible to TNF-induced cytotoxicity [62]. However, TRAF2 cannot bind to E2 enzymes like Ubc13, as the structure of its RING domain is different from the TRAF6 RING domain [64] and it cannot catalyze the synthesis of K63-linked ubiquitin chains. Hence, it can be interpreted that the important role of TRAF2/5 in the TNFα-mediated pathway is a consequence of its adaptor function. Of course, it is possible that the adaptor function of TRAF2 acts to recruit another ligase (like cIAP) to RIP1.

In order to ligate ubiquitin chains on to RIP1, both the TRAFs and cIAPs would need to collaborate with the E2 enzyme UBC13. It has been reported that a dominant–negative mutant of UBC13 blocks TNFα and TRAF2-mediated NF-κB activation [3]. Ubc13 knockouts are embryonic lethal; however it is not entirely clear if this is because of dysregulation of NF-κB signaling [18, 65]. There is conflicting genetic data on this topic: while one group showed reduced activation of NF-κB in UBC13$^{+/-}$ macrophages and splenocytes [65], another group demonstrated that UBC13$^{-/-}$ MEFs have normal NF-κB signaling in response to TNFα [18]. These discrepancies may be attributed to the different cell lines used by the researchers or to redundancy with another E2 enzyme. In this regard, it has been shown that another E2 enzyme (UBC4/5) can also activate IKK (the ligase is as yet unidentified) [2]. It is safe to conclude that much more work needs to be done to identify the physiologically relevant E2 enzyme and E3 ligase, if K63-linked regulatory ubiquitination indeed plays a central role in the NF-κB pathway.

In a more recent finding, another E3 ligase which catalyzes the formation of linear ubiquitin chains [66] has been shown to be recruited to the TNFR1 upon receptor stimulation [67]. This E3 ligase complex is known as LUBAC and is composed of HOIP, HOIL-1, and Sharpin [68]. The E3 ligase activity has been verified in vitro as LUBAC and Ubc5 have been shown to add linear polyubiquitin chains to NEMO [68]. This linear polyubiquitination of NEMO has been proposed to be important for NF-κB activation. However, the recruitment of LUBAC to TNFR requires only the TRADD, TRAF2, and cIAP1/2 proteins. It does not require NEMO or RIP1, thus it is difficult to understand how linear ubiquitination of NEMO might activate NF-κB in a cellular setting. Indeed, there is no evidence of linearly ubiquitinated NEMO in cells.

Additionally, RNAi of HOIP only partially inhibits TNFα-induced activation of IKK. A naturally occurring mutant mouse (called cpdm) that has a stop codon in the sharpin gene and is deficient for sharpin expression shows chronic proliferative dermatitis and reduced IKK activation [69–71]. Surprisingly though, others reported that these mice have normal IKK or NF-κB activation (the data even shows a somewhat enhanced IKK activity) [72]. Even if these mice were to show an inflammatory phenotype, it clearly does not resemble the lethality that is associated with knockouts of IKK components [73]. Crossing the cpdm mice with the IL-1 receptor accessory protein (IL-1RAcP)-deficient mice rescues the inflammatory phenotype;

thus the inflammation is caused by enhanced IL-1β signaling [74]. Treatment of cpdm mice with a proteasome inhibitor like bortezomib also ameliorates the inflammation, which is further evidence that the inflammation in these mice results from excessive NF-κB activation (and not decreased NF-κB activation as the model would suggest). Other factors might also be involved in the cpdm mice as Sharpin is also known to inhibit activation of integrins [75].

While the knockouts of the various components that mediate regulatory ubiquitination do not all exhibit phenotypes expected to result from knocking out essential member of the NF-κB pathway, genetic ablation of A20 indeed resulted in a drastic phenotype and mice that die from multiorgan inflammation and cachexia [37]. This is because A20 knockouts show persistent and hyperactivated NF-κB signaling. A20 has a unique mechanism of action; it has been shown to first remove K63-linked polyubiquitin chains from RIP1 to attenuate NF-κB signaling, and then ligate K48-linked polyubiquitin chains on RIP1 to completely inhibit NF-κB [63]. As I will show in my thesis, the deubiquitinase activity of A20 is actually dispensable in the physiological context. This is quite remarkable given the wealth of biochemical data stating otherwise. A number of other deubiquitinases of RIP1 have also come to light. A prominent example is CYLD which deubiquitinates RIP1 and inhibits NF-κB in response to TNF [33]. Another DUB is Cezanne which resembles A20, has an OTU domain, and deubiquitinates RIP1 upon TNFα stimulation [76]. Figure 2.2 shows how regulatory ubiquitination could potentially regulate the NF-κB pathway downstream of TNFR.

2.4 Deubiquitinases and NF-κB

The process of ubiquitination can be reversed by the action of deubiquitinases (DUBs) just like phosphorylation can be reversed by phosphatases. Thus, deubiquitinases can reverse the effect of E3 ligases. Bioinformatic estimates predict around 100 deubiquitinases in the genome [77, 78]. While some DUBs are metalloproteases, most are cysteine proteases. The cysteine proteases may be classified by the presence of distinct protease domains. They may be ovarian tumor (OTU) proteases, ubiquitin-specific proteases (USP), Machado-Joseph disease proteases (MJD), or ubiquitin C-terminal hydrolases (UCH).

A20 and CYLD are the common deubiquitinases in the NF-κB pathway. While A20 is a OTU deubiquitinase, CYLD is a USP deubiquitinase. Nevertheless, they share many of their substrates. Since A20-knockouts die from multiorgan inflammation and cachexia, it is clear there is no functional redundancy with CYLD. This is probably because of temporal reasons, as they target the substrates at two different times during the inflammatory process [79]. CYLD prevents spontaneous activation of NF-κB in quiescent cells while A20 is essential to terminate NF-κB signaling in activated cells.

CYLD is constitutively active and is inactivated in stimulated cells by IKK-mediated phosphorylation (in a NEMO-dependent manner) [80]. However, it is unclear from the data if the transient phosphorylation happens before TRAF2 ubiquitination and NF-κB activation, as would be expected if CYLD was regulating the process [80]. A20 on the other hand is basally expressed at very low levels but rapidly induced by proinflammatory stimuli to downregulate NF-κB after an initial inflammatory phase [81]. Unlike CYLD, A20 is a NF-κB-dependent gene. While expression of A20 is mainly controlled by NF-κB-mediated transcription, it is also posttranslationally modified by IKKβ-induced phosphorylation at Ser381 [82]. Phosphorylation of A20 increases its ability to downregulate NF-κB. Considered together, phosphorylation and inactivation of CYLD by IKK provides a window for NF-κB activation before it is inducibly turned off by A20 by a negative feedback mechanism. Hence A20 and CYLD serve precise and temporally distinct roles in downregulating NF-κB [81].

It is worth noting that while the proposed deubiquitinase activity of both A20 and CYLD is predicated on K63-linked polyubiquitin chains, they both promiscuously cleave other chains as well. CYLD has been shown to deubiquitinate linear polyubiquitin chains (besides K63-linked polyubiquitin chains) [83]. For specificity, CYLD has been proposed to require various adaptor molecules like p62 in osteoclasts (selectively binds to and recruits CYLD to deubiquitinate TRAF6 downstream of RANK) [84]. In vitro, A20 has been shown to have a greater reactivity towards K48-linked polyubiquitin chains and deubiquitinates K48-linked chains better than K63-linked polyubiquitin chains [85, 86]. To achieve this preference for K63-linked chains in vivo, A20 forms a complex with other proteins like TAX1BP1 [87].

Just as there is strong biochemical evidence implicating E3 ligases in NF-κB activation, there is also similar evidence that implicates deubiquitinating enzymes in the downregulation of NF-κB. Additionally, complete genetic knockouts show that deletion of these DUBs indeed lead to rampant inflammation. Thus, both CYLD and A20 are very important in negatively regulating NF-κB activation. However, to conclusively prove the physiological requirement of the deubiquitinase function, mice containing knock-in mutations which destroy the deubiquitinase function would be necessary. Given the clear physiological importance of A20 and CYLD, it is essential to decipher the mechanism of action for these two proteins.

Since this thesis studies the mechanism of action of A20, this deubiquitinase is discussed in detail below.

2.5 A20 (TNFAIP3 or TNFα-Induced Protein 3 Gene)

2.5.1 Structure and Proposed Mechanism of Action

A20 was discovered in human umbilical vein endothelial cells (HUVEC) as a zinc-finger protein that is rapidly induced upon stimulation with TNFα, hence it is also called "TNFα-Induced Protein 3" [88, 89]. The function of A20 was to provide protection from TNF-mediated cytotoxicity [89]. It was shown that A20 is virtually absent at basal levels, but rapidly induced by a plethora of PAMPS, mitogens, or

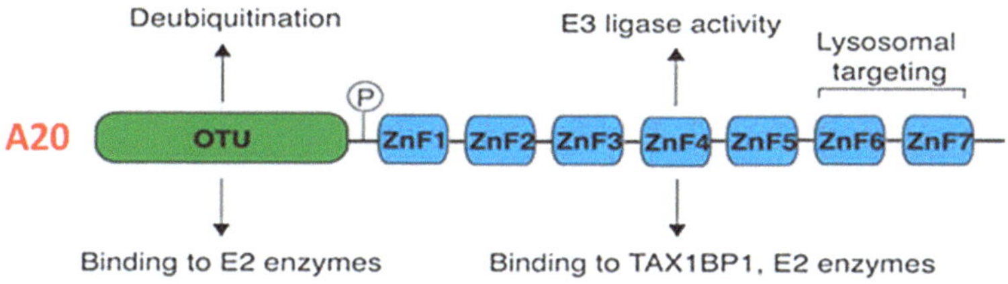

Fig. 2.3 Structure of A20 (Reprinted by permission from Macmillan Publishers Ltd: [Cell Research] (Harhaj EW, Dixit VM (2011) Deubiquitinases in the regulation of NF-[kappa]B signaling. Cell Res 21: 22-39), copyright (2011))

proinflammatory cytokines [88] in a NF-κB-dependent manner. While cells of the innate immune system (dendritic cells, macrophages) show inducible expression of A20 [88], thymocytes have constitutive expression of high levels of A20 that is downregulated upon TCR stimulation [79, 90].

An analysis of the structure of A20 (Fig. 2.3) shows a N-terminal OTU domain followed by seven C-terminal zinc-finger domains. A cysteine residue at position 103 (in the OTU domain) has been shown to be important for the deubiquitinase activity [63, 91]. Multiple studies based on overexpression have demonstrated that A20 downregulates NF-κB signaling downstream of the TNFR and TLRs [35, 92, 93]. Upon TNFR stimulation, A20 is inducibly expressed and targets ubiquitinated RIP1 in an enigmatic two-step sequential process to downregulate NF-κB [63]. A20 acts as a dual ubiquitin-editing enzyme that can act both as a deubiquitinase and an E3 ligase. The Cys103 residue in the OTU domain first deubiquitinates K63-linked polyubiquitin chains from RIP1, thereby attenuating NF-κB activation. After this, the Cys624/Cys627 in the fourth zinc-finger domain (ZNF4) ligates K48-linked polyubiquitin chains to RIP1, thereby triggering proteasomal degradation of RIP1 and completely inactivating NF-κB [63]. E3 ligases are typically characterized by the presence of a RING or HECT domain; however, the E3 ligase function of A20 is uniquely mediated by a zinc-finger domain. A considerable fraction of cellular A20 has been found to localize to lysosomes. This localization is mediated by zinc-finger domains and has been found to lead to degradation of TRAF2 [94, 95]. Surprisingly though, a deubiquitinase mutant of A20 (C103A) has been shown to inhibit IKK both in vitro and in vivo, leading to normal activation of NF-κB [96].

A20 can also inhibit NF-κB in response to LPS stimulation. However, it does so by a different mechanism. Once induced, A20 attenuates NF-κB activation by preventing further K63-linked ubiquitination by disrupting interactions between Ubc13 (E2 enzyme) and TRAF6 (E3 ligase) [34]. At later time points, A20 abrogates NF-κB signaling by adding K48-linked polyubiquitin chains to Ubc13 targeting it for proteasomal degradation. This process is also dependent on ZNF4 (C624/C627) and Cys103. Another publication showed that A20 directly inhibited LPS-mediated NF-κB signaling by deubiquitinating TRAF6 [97].

The mechanisms are elegant and C624/C627 and C103 have been shown to be important in downregulating NF-κB both in response to TNF and LPS stimulation.

However the experiments are based on overexpression and knockdown of the protein. This thesis will characterize knock-in mice in which the Cys103 residue is mutated to test the physiological importance of the deubiquitinase function of A20.

2.5.2 Function of A20

The antiapoptotic and NF-κB inhibitory role of A20 was elucidated after the generation of A20-knockout mice. A20-knockouts are perinatally lethal. They die because of spontaneous, extensive multiorgan inflammation and cachexia, resulting from persistent, unabated NF-κB signaling [37]. As would be expected, they also succumb to different proinflammatory stimuli including sub-lethal doses of TNF-α or LPS [37]. A20-knockout MEFs show persistent activation of NF-κB in response to TNF-α, LPS, and IL-1 stimulation and are more sensitive to apoptosis as compared to wild-type MEFs [37].

The inflammation in A20-knockouts is mediated by myeloid cells and is independent of the adaptive response. This was found by deleting A20 in mice lacking B and T cells (*Rag1*$^{-/-}$ mouse). Thus, inflammation is not rescued in a *TNFAIP3*$^{-/-}$*Rag1*$^{-/-}$ double deficient mice [37]. Surprisingly, even though A20 was considered a feedback inhibitor of TNF-mediated signaling, the inflammation in A20-knockouts is independent of TNFR signaling. Thus the inflammation also continues unabated in *TNFAIP3*$^{-/-}$ *TNFα*$^{-/-}$ and *TNFAIP3*$^{-/-}$ *TNFR1*$^{-/-}$ double-knockouts [97]. While the deubiquitinating property of A20 is certainly important in vitro, this observation casts serious doubts on the physiological relevance of this function. This is because the deubiquitinase role has been implicated in downregulating TNF signaling, and the physiological relevance of A20 is clearly not dependent on TNF signaling.

The inflammation in A20 knockouts is recued by crossing with *Myd88*$^{-/-}$ mice. Thus, the spontaneous inflammation in A20 knockouts is triggered by TLR-mediated signaling, and the *TNFAIP3*$^{-/-}$ *Myd88*$^{-/-}$ double-knockouts are normal [97, 98]. Consistently, A20-deficient mice can be successfully treated with broad-spectrum antibiotics and cured of the severe inflammation [98]. Hence, it can be inferred that A20-knockouts have constitutive TLR signaling driven by commensal intestinal flora. Thus, A20 is essential to maintain intestinal immune homeostasis.

A20 has also been specifically deleted from various organs and these conditional-knockouts have been used to determine the role of A20 in these organs (Table 2.1). Thus, A20 was specifically deleted from intestinal epithelial cells (IECs) by crossing with Villin-Cre mice [99]. The A20 IEC conditional knockout (A20^{IEC-KO}) mice were found to be healthy. However, they have a greater propensity of developing colitis when induced with dextran sulfate sodium (DSS). This colitis is rescued by crossing with TNFR knockouts [99]. Thus, TNFR signaling is responsible for the systemic, intestinal pathophysiology.

Table 2.1 Phenotypes of cell type-specific deletion of A20

Cell type	Genetic modification	Mouse phenotype	Related human disease
B cells	Cre-mediated deletion *of TNFaip3* in cells that express CD19	Germinal centre and plasma cell dysplasis; production of autoantibodies; renal immunoglobulin deposition; B cell resistance to FAS-mediated cell death	Systemic lupus erythematosus
DCs	Cre-mediated deletion of *TNFaip3* in cells that express CD 11c	DC activation; expansion and activation of T cell and myeloid cell populations; colitis; spondyloarthritis	Inflammatory bowel disease
DCs	Cre-mediated deletion of *TNFaip3* in cells that express CD	DC activation; expansion of T cell and plasma cell populations; increased uptake of apoptotic cells by DCs; autoantibody production; nephritis	Systemic lupus erythematosus
Macrophages and granulocytes	Cre-mediated deletion of *TNFaip3* in cells that express lysozyme M	Increase IL-6 production; production of collagen-specific autoantibodies; protection against influenza A virus infection	Rheumatoid arthritis
Intestinal epithelial cells	Cre-mediated deletion of *TNFaip3* in cells that express villin	Hypersensitivity to experimental colitis	Inflammatory bowel disease
Intestinal epithelial cells	Villin-driven expression of a *TNFaip3* transgene	Protection against DSS-induced colitis	Inflammatory bowel disease
Keratinocytes	Cre-mediated deletion of *TNFaip3* in cells that express keratin 14	Epidermal hyperproliferation; hair and skin defects; sebaceous gland hyperplasia	?

Adopted by permission from Macmillan Publishers Ltd: [Nature Reviews Immunology] (Ma A, Malynn BA (2012) A20: linking a complex regulator of ubiquitylation to immunity and human disease. Nat Rev Immunol 12: 774-785), copyright (2012)

A20 plays a profound role in the innate immune system. Specific ablation of A20 in myeloid cells causes spontaneous development of rheumatoid arthritis accompanied by osteoclastogenesis [100]. This is caused by a higher level of proinflammatory serum cytokines, as would be expected from protracted activation of NF-κB [100]. The pathology was shown to be dependent on TLR4-MyD88 and

IL-6 signaling but independent of TNF [100]. Upon stimulation with LPS, both macrophages [97] and dendritic cells (DCs) [101] deficient in A20 produced more proinflammatory cytokines (IL-6 and TNF-α) and showed an enhanced expression of co-stimulatory molecules [101]. siRNA-based silencing of A20 in DCs triggered constitutive NF-κB activation and antigen presentation [101]. Thus, A20 seemed to be important in not only determining the activation threshold of DCs, but also in regulating antigen presentation. Indeed when A20 is selectively deleted from DCs by crossing with CD11-Cre transgenic mice, the mice develop autoimmunity and show spontaneous proliferation of T cells and expansion of plasma cells. They have symptoms of systemic lupus erythematosus (SLE) [102]. The DCs were resistant to apoptosis probably because of upregulation of Bcl-2 and Bcl-x [102]. In this regard, patients of SLE [103] have multiple SNPs associated with A20.

The importance of A20 in mediating antigen presentation of DCs might imply that A20 may also be important for adaptive immunity. In fact, A20 has been determined to be essential for terminating NF-κB signaling in both T and B lymphocytes. The expression pattern of A20 in these lymphoid cells is different from myeloid cells. In lymphoid cells, A20 is constitutively expressed. The basal levels of A20 are particularly high in T cells [90]. Experiments based on overexpression reveal that A20, but not the deubiquitinase mutant, can downregulate NF-κB signaling in response to CD3 and CD28 receptor stimulation [104]. A20 acts by removing K63-linked polyubiquitin chains from MALT1 to downregulate NF-κB [105]. Conversely MALT1 has paracaspase activity to cleave A20 and facilitate NF-κB activation [106]. This balance between "MALT1-mediated A20 cleavage" and "A20-mediated MALT1 deubiquitination" finetunes NF-κB activity in T cells.

A20 is constitutively expressed in B lymphocytes as well. Intriguingly though, CD40 signaling further increases the expression of A20 [107]. This would suggest that A20 plays an important role in downregulating NF-κB in these cells. Indeed, selectively deleting A20 in B cells (by crossing with CD19-Cre mice) results in autoimmunity mediated by production of autoantibodies [108]. The ($A20^{CD19\text{-}KO}$) mice show normal development and growth. However, they have more immature and germinal center (GC) B cells as compared to their littermates. Their splenic B cells also proliferate more upon stimulation with α-CD40, LPS, or CpG [108]. They show enhanced non-canonical and canonical NF-κB signaling as evidenced by phosphorylation of p100 and degradation of IκBα respectively [108]. Thus, A20 maintains B-cell homeostasis by inhibiting both canonical and non-canonical signaling in B lymphocytes. However, one observation in these mice ($A20^{CD19\text{-}KO}$) does not correlate with the well-documented antiapoptotic effects of A20. The B cells in these mice are surprisingly resistant to Fas-mediated apoptosis. This is because of the higher expression of antiapoptotic molecules like Bcl-x in these mice.

The cross-regulation of A20 and MALT1 is not unique to T cells alone, but is present in B cells as well. Dysregulation of this cross-regulation may result in B-cell lymphoma. In B-cell lymphomas, the paracaspase activity of MALT1 is constitutively active. This deactivates A20, and leads to chronic, persistent activation of NF-κB resulting in lymphoma [109].

2.5.3 Regulation of A20 by Ubiquitin-Binding Proteins and Adaptors

A20 exists in a complex with many other proteins like Tax1-binding protein 1 (TAX1BP1), Itch, Ring finger protein (RNF) 11, and possibly ABIN-1 and YMER, and this "A20 ubiquitin-editing complex" has been proposed to be important for inhibiting NF-κB signaling [79]. Thus, the interaction of A20 with RIP1 or TRAF6 is mediated by TAX1BP1 [110] while another E3 ligase (Itch) is essential for A20 promoting K48-linked polyubiquitination of RIP1 [111]. ABIN-1 has been shown to be important in terminating the NF-κB response by deubiquitinating NEMO [112, 113].

TAX1BP1 was discovered using the HTLV-I Tax oncoprotein as bait in a yeast two-hybrid screen [114]. Later, TAX1BP1 was shown to bind A20 and cooperatively promote cell survival [87]. TAX1BP1 knockouts have been generated using two methods, by gene trapping and by a conventional gene targeting method. The resulting phenotypes are different for unknown reasons. The knockout generated using gene trapping is embryonically lethal because of cardiac defects at E13.5 [115]. The knockout generated using gene targeting [110] developed normally till 4 months. After this, they succumbed to inflammatory cardiac valvulitis. While the first phenotype resembles the A20-knockout and suggests that TAX1BP1 could be an essential "adaptor protein for A20," the second phenotype does not readily comply with the mechanism of cooperative-action.

However both the A20 and the TAX1BP1 knockouts are hypersensitive to proinflammatory cytokines because of enhanced activation of NF-κB [110, 115]. They also succumb to sub-lethal doses of IL-1β, and TNF-α. RIP1 is persistently ubiquitinated in *Tax1bp1*$^{-/-}$ MEFs upon TNF stimulation, similar to the A20-knockout MEFs [115]. This suggests that A20 and TAX1BP1 act in a complex to reduce RIP1 ubiquitination. Since TAX1BP1 does not have deubiquitinating activity, it probably inhibits RIP1 ubiquitinating by acting through A20 [115].

TAX1BP1 has two zinc-finger (ZnF) domains in the C-terminus and one of them is known to harbor a UBD. This UBD recognizes the K63-linked polyubiquitin chains of RIP1, serving as the adaptor domain for TAX1BP1 to recruit A20 to the receptor complex [110, 115]. The ZnF domains of a TAX1BP1 have a "PPXY" motif (P-Proline, X-variable amino acid, Y-tyrosine). Such motifs preferably interact with "WW" (W-tryptophan) motifs in other proteins [116]. This motif is important for a functional TAX1BP1 molecule as TAX1BP1 mutants lacking this motif have persistent NF-κB activation [111]. It was later discovered that Itch contains the "WW" motif and interacts with TAX1BP1 [111]. The TAX1BP1/Itch complex is inducibly formed upon TNF stimulation and is responsible for recruiting A20 to the receptor complex and deubiquitinating RIP1 [111].

Itch-knockouts also show hyperactive NF-κB signaling, however the inflammation in these mice is confined to the lungs and skin alone [117] as compared to A20-knockouts, which show inflammation throughout the body. Additionally, unlike A20—knockouts, the inflammation in Itch-deficient mice depends on the adaptive immune system and is recued by crossing with *Rag1*$^{-/-}$ mice [111]. Thus, it is probable that Itch has functions independent of A20.

RNF11 also interacts with A20, TAX1BP1, Itch, and NEMO as identified in a high-throughput yeast two-hybrid screen [118]. RNF11 is an E3 ligase and is overexpressed in cancers of different tissues (breast, pancreas, etc) [119, 120]. This suggested a possible link with the NF-κB pathway and it was subsequently demonstrated that RFN11–TAX1BL1–Itch complex is inducibly formed upon stimulation with TNF-α or IL-1 [121]. Overexpression of RNF11 negatively regulates NF-κB signaling, while knockdown of RNF11 using siRNA increases activation of NF-κB [121]. Furthermore, cells deficient in RFN11 showed increased ubiquitination of RIP1 and TRAF6 [121]. Hence, it has been suggested that A20 acts in a complex with RFN11, TAX1BL1, and Itch. RNF11 is also known to participate in TGF-β signaling [122, 123]. Hence, RNF11 clearly plays a complex role and delineation of its physiological function will have to await the generation of knockout mice.

ABIN-1 is another A20-interacting protein identified by a yeast two-hybrid screen [93], whose overexpression inhibits NF-κB signaling in response to a wide range of stimuli [93, 124]. ABIN1 has been proposed to be an adaptor molecule essential for the interaction of A20 and NEMO, leading to deubiquitination of NEMO by A20 [113]. Interestingly, a pathogenic E3 ligase IpaH9.8 found in Shigella cooperates with ABIN-1 to add K48-linked polyubiquitin chains to NEMO, thereby targeting it for proteasomal degradation and inactivation NF-κB in the process [125]. However, a major discrepancy is that MEFs deficient in ABIN-1 show normal NF-κB activation [126].

Similarly, the yeast two-hybrid screen identified another A20 interaction protein named YMER [127]. YMER was also identified in a protein array using a polyubiquitin bait [128], and this polyubiquitin-binding domain of YMER is essential for inhibition of NF-κB signaling [127]. Overexpressing YMER downregulates NF-κB, while YMER knockdown cells show augmented NF-κB activation [127]. YMER interacts with RIP1 and hence has been proposed to act in a complex with A20 and aid in RIP1 deubiquitination [127].

2.5.4 The Role of A20 in Autoimmune Diseases and Cancer

Polymorphisms of the TNFAIP3 gene (Fig. 2.4) has been implicated in a number of autoimmune diseases (rheumatoid arthritis, type 1 diabetes, inflammatory bowel disease, psoriasis, SLE, coronary artery disease, celiac disease, and SLE) [129–138]. Based on murine studies, it could be rationalized that these polymorphisms affect the expression and/or function of the expressed A20 protein [79]. A mutation in the coding region of the OTU domain (A125V) has been found in the African population [139]. The data tentatively suggests that this affects A20-mediated deubiquitination of TRAF2. Two other mutations (V377M and P656L) are associated with lung cancers. The mechanistic effect of these two mutations on A20 is not clear.

A20 acts as a tumor suppressor for many different B-cell lymphomas. A20 is on chromosome 6, and deletion of chromosome 6q is often associated with non-

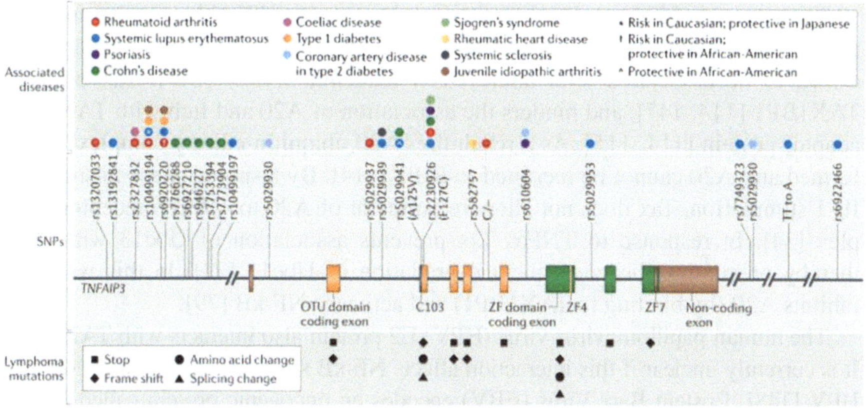

Fig. 2.4 Polymorphisms in A20 associated with human diseases (Reprinted by permission from Macmillan Publishers Ltd: [Nature Reviews Immunology] (Ma A, Malynn BA (2012) A20: linking a complex regulator of ubiquitylation to immunity and human disease. Nat Rev Immunol 12: 774-785), copyright (2012))

Hodgkin lymphomas [140]. Several point mutations that could inactivate A20 lead to marginal zone lymphoma, Hodgkin lymphoma, diffuse large B-cell lymphoma (DLBCL), or MALT lymphoma [130, 131, 133, 141]. A reduction in the levels of the A20 protein (by methylation of promoter) could lead to MALT lymphoma [142]. The tumor suppressor function of A20 has been conclusively proven by demonstrating that B-cell lymphomas can be rescued by expressing A20 in lymphoma cell lines that lacks both A20 alleles [130, 131].

It is interesting to note that oncogenic transformation of MEFs has been found to be associated with reduction of expression of A20 protein. For example, Ras-mediated transformation of primary MEFs leads to the upregulation of most NF-κB-controlled genes. It, however, reduces levels of A20 [143].

A20 inhibits TNF-mediated cytotoxicity [89] and has prosurvival activity in this context. Hence, it is conceivable that A20 may be an oncogene in certain circumstances. This is indeed the case in breast cancer and glioma. A20 is estrogen-regulated in breast cancer and protects against the apoptotic effects of tamoxifen [144]. A20 is needed for glioma stem cell survival [145]. Indeed overexpression of A20 is a hallmark of breast tumor and glioma and may serve as a therapeutic marker for these diseases [144, 145].

2.5.5 *Pathogens Modulating A20 Activity*

Since A20 is an essential regulator of NF-κB, oncogenic viruses often inactivate it, thus promoting chronic activation of NF-κB [79]. A few examples are outlined below. They highlight the critical role that A20 plays in controlling excessive immune responses.

The genome of Human T-cell leukemia virus type I (HTLV-I) encodes an oncogenic protein called Tax which mediates activation of NF-κB and has been hypothesized to be associated with adult T-cell leukemia (ATL) [146]. Tax binds to TAX1BP1 [114, 147], and hinders the association of A20 and Itch with TAX1BP1 adaptor protein [114, 115]. As a result the "A20 ubiquitin-editing complex" is not formed and A20 cannot be recruited to RIP1 [114]. By a similar mechanism, upon IL-1 stimulation, Tax does not allow recruitment of A20 to TRAF6 receptor complex [34]. In response to TNFα, Tax prevents association of Ubc13 with A20, thereby preventing the consequent degradation of Ubc13 [34]. In this way, Tax inhibits A20 (by binding to TAX1BP1) and activates NF-κB [79].

The human papillomavirus virus (HPV) E2 protein also interacts with TAX1BP1. It is currently unclear if this interaction affects NF-κB signaling during infection with HPV [148]. Epstein Barr Virus (EBV) encodes an oncogenic protein called LMP1. This protein interacts with A20 to promote constitutive activation of NF-κB [149, 150]. A20 also deubiquitinates IRF7 in Raji cells infected with EBV virus [151]. Thus, the infection is not mediated solely through the inactivation of the NF-κB pathway.

In addition to encoding for proteins that hijack the cellular role of A20, some viruses and bacteria have also evolved to directly encode for a DUB that resembles A20. This could in theory allow the pathogen to inhibit NF-κB [79] and reduce the proinflammatory response by the host. In effect, this would provide the pathogen a strategy for "immune evasion." The Crimean Congo hemorrhagic fever virus (CCHFV) causes hemorrhages and 30 % of infections are lethal [152]. It encodes a protein named L protein which has a N-terminal OTU domain (just like A20). This is not unique to CCHFV, but also found in Nairobi sheep disease virus and nairoviruses Dugbe virus (DUGV). This protein deubiquitinates NF-κB activators, thereby attenuating TNF-α-mediated activation of NF-κB. The bacteria Yersinia (causes plague) has evolved to encode a gene called YopJ. This virulence factor could potentially deubiquitinate TRAF2 or TRAF6, leading to inactivation of NF-κB and reduced expression of proinflammatory cytokines [153]. Much like mammalian DUBs, pathogenic DUBs also seem to lack specificity. Thus, YopJ has been found to cleave both K63-linked and K48-linked polyubiquitin chains [153]. In this way, by mimicking the DUBs, pathogenic bacteria and viruses can find a way to escape the host immune response.

2.5.6 Regulation of A20 Deubiquitinase Activity

Recently, the crystal structure of the OTU domain was solved independently by two groups [86, 154]. The structure is similar to other cysteine proteases, and the architecture of catalytic cysteine (C103) and accompanying histidines is conserved [154]. The C103 (responsible for deubiquitinase function) is located in an alpha helix, which may provide the platform for binding of ubiquitin chains [86]. However biochemical evidences by other groups have suggested that the seventh zinc-finger motif is important for ubiquitin binding [96].

A20 cleaves K48-linked polyubiquitin chains rather than K63-linked polyubiquitin chains in vitro, and this is at odds with the proposed mechanism of disassembling K63-linked chains from RIP1 and TRAF6 [85, 86]. Additionally, unlike most of cysteine proteases of OTU family, A20 deubiquitinates K63-linked chains from the TRAF6/ubiquitin interface [86]. While some groups have confirmed the importance of the deubiquitinase activity of A20 [63, 97, 115], others have found that that this property of A20 is dispensable for downregulating NF-κB [91, 94]. Moreover, inhibition of antiviral signaling requires the ZF domains [155] but not the DUB domain of A20 [155–157]. These discrepancies may be attributed to distinct experimental designs employed by different groups (cell lines, varying levels of overexpression, etc).

2.6 Conclusion and Perspectives

Despite the large body of literature implicating A20 in inflammatory responses, the mechanism by which A20 downregulates inflammation has remained elusive. Multiple biochemical and knockdown approaches in cell culture have suggested that A20 functions as a deubiquitinase by disassembling regulatory K63-ubiquitin chains on upstream signaling molecules such as RIP1. In the next chapter, we report the creation and characterization of a knock-in mouse that expresses a mutated form of A20 that lacks the deubiquitinase activity. The knock-in mouse that we have generated abrogates the deubiquitinase function of A20 in vivo, and is the ideal tool to study the physiological role of the deubiquitinase function of A20.

References

1. Chen ZJ (2005) Ubiquitin signalling in the NF-kappaB pathway. Nat Cell Biol 7:758–765
2. Chen ZJ, Parent L, Maniatis T (1996) Site-specific phosphorylation of IkappaBalpha by a novel ubiquitination-dependent protein kinase activity. Cell 84:853–862
3. Deng L, Wang C, Spencer E, Yang L, Braun A, You J, Slaughter C, Pickart C, Chen ZJ (2000) Activation of the IkappaB kinase complex by TRAF6 requires a dimeric ubiquitin-conjugating enzyme complex and a unique polyubiquitin chain. Cell 103:351–361
4. Ishida T et al (1996) Identification of TRAF6, a novel tumor necrosis factor receptor-associated factor protein that mediates signaling from an amino-terminal domain of the CD40 cytoplasmic region. J Biol Chem 271:28745–28748
5. Cao Z, Xiong J, Takeuchi M, Kurama T, Goeddel DV (1996) TRAF6 is a signal transducer for interleukin-1. Nature 383:443–446
6. Kobayashi N, Kadono Y, Naito A, Matsumoto K, Yamamoto T, Tanaka S, Inoue J (2001) Segregation of TRAF6-mediated signaling pathways clarifies its role in osteoclastogenesis. EMBO J 20:1271–1280
7. Walsh MC, Kim GK, Maurizio PL, Molnar EE, Choi Y (2008) TRAF6 autoubiquitination-independent activation of the NFkappaB and MAPK pathways in response to IL-1 and RANKL. PLoS One 3:e4064

8. Liu S, Chen ZJ (2011) Expanding role of ubiquitination in NF-[kappa]B signaling. Cell Res 21:6–21
9. Wertz IE, Dixit VM (2010) Signaling to NF-kappaB: regulation by ubiquitination. Cold Spring Harb Perspect Biol 2:a003350
10. Chen J, Chen ZJ (2013) Regulation of NF-kappaB by ubiquitination. Curr Opin Immunol 25:4–12
11. Zhou H, Wertz I, O'Rourke K, Ultsch M, Seshagiri S, Eby M, Xiao W, Dixit VM (2004) Bcl10 activates the NF-kappaB pathway through ubiquitination of NEMO. Nature 427:167–171
12. Ni CY et al (2008) Cutting edge: K63-linked polyubiquitination of NEMO modulates TLR signaling and inflammation in vivo. J Immunol 180:7107–7111
13. Dunne A, O'Neill LA (2003) The interleukin-1 receptor/Toll-like receptor superfamily: signal transduction during inflammation and host defense. Sci STKE 171:re3
14. Yamaguchi K, Shirakabe K, Shibuya H, Irie K, Oishi I, Ueno N, Taniguchi T, Nishida E, Matsumoto K (1995) Identification of a member of the MAPKKK family as a potential mediator of TGF-beta signal transduction. Science 270:2008–2011
15. Wang C, Deng L, Hong M, Akkaraju GR, Inoue J, Chen ZJ (2001) TAK1 is a ubiquitin-dependent kinase of MKK and IKK. Nature 412:346–351
16. Kanayama A, Seth RB, Sun L, Ea CK, Hong M, Shaito A, Chiu YH, Deng L, Chen ZJ (2004) TAB2 and TAB3 activate the NF-kappaB pathway through binding to polyubiquitin chains. Mol Cell 15:535–548
17. Shim JH et al (2005) TAK1, but not TAB1 or TAB2, plays an essential role in multiple signaling pathways in vivo. Genes Dev 19:2668–2681
18. Yamamoto M et al (2006) Key function for the Ubc13 E2 ubiquitin-conjugating enzyme in immune receptor signaling. Nat Immunol 7:962–970
19. Xu M, Skaug B, Zeng W, Chen ZJ (2009) A ubiquitin replacement strategy in human cells reveals distinct mechanisms of IKK activation by TNFalpha and IL-1beta. Mol Cell 36:302–314
20. Yamamoto M, Sato S, Saitoh T, Sakurai H, Uematsu S, Kawai T, Ishii KJ, Takeuchi O, Akira S (2006) Cutting edge: pivotal function of Ubc13 in thymocyte TCR signaling. J Immunol 177:7520–7524
21. Yamazaki K et al (2009) Two mechanistically and temporally distinct NF-kappaB activation pathways in IL-1 signaling. Sci Signal 2(93):ra66
22. Sayama K et al (2010) E2 Polyubiquitin-conjugating enzyme Ubc13 in keratinocytes is essential for epidermal integrity. J Biol Chem 285:30042–30049
23. Conze DB, Wu CJ, Thomas JA, Landstrom A, Ashwell JD (2008) Lys63-linked polyubiquitination of IRAK-1 is required for interleukin-1 receptor- and toll-like receptor-mediated NF-kappaB activation. Mol Cell Biol 28:3538–3547
24. Sun L, Deng L, Ea CK, Xia ZP, Chen ZJ (2004) The TRAF6 ubiquitin ligase and TAK1 kinase mediate IKK activation by BCL10 and MALT1 in T lymphocytes. Mol Cell 14:289–301
25. Windheim M, Stafford M, Peggie M, Cohen P (2008) Interleukin-1 (IL-1) induces the Lys63-linked polyubiquitination of IL-1 receptor-associated kinase 1 to facilitate NEMO binding and the activation of IkappaBalpha kinase. Mol Cell Biol 28:1783–1791
26. Trompouki E, Hatzivassiliou E, Tsichritzis T, Farmer H, Ashworth A, Mosialos G (2003) CYLD is a deubiquitinating enzyme that negatively regulates NF-kappaB activation by TNFR family members. Nature 424:793–796
27. Yin Q et al (2009) E2 interaction and dimerization in the crystal structure of TRAF6. Nat Struct Mol Biol 16:658–666
28. Lamothe B, Webster WK, Gopinathan A, Besse A, Campos AD, Darnay BG (2007) TRAF6 ubiquitin ligase is essential for RANKL signaling and osteoclast differentiation. Biochem Biophys Res Commun 359:1044–1049
29. Lamothe B, Besse A, Campos AD, Webster WK, Wu H, Darnay BG (2007) Site-specific Lys-63-linked tumor necrosis factor receptor-associated factor 6 auto-ubiquitination is a critical determinant of I kappa B kinase activation. J Biol Chem 282:4102–4112

30. Xia ZP, Sun L, Chen X, Pineda G, Jiang X, Adhikari A, Zeng W, Chen ZJ (2009) Direct activation of protein kinases by unanchored polyubiquitin chains. Nature 461:114–119
31. Kovalenko A, Chable-Bessia C, Cantarella G, Israel A, Wallach D, Courtois G (2003) The tumour suppressor CYLD negatively regulates NF-kappaB signalling by deubiquitination. Nature 424:801–805
32. Brummelkamp TR, Nijman SM, Dirac AM, Bernards R (2003) Loss of the cylindromatosis tumour suppressor inhibits apoptosis by activating NF-kappaB. Nature 424:797–801
33. Sun SC (2010) CYLD: a tumor suppressor deubiquitinase regulating NF-kappaB activation and diverse biological processes. Cell Death Differ 17:25–34
34. Shembade N, Ma A, Harhaj EW (2010) Inhibition of NF-kappaB signaling by A20 through disruption of ubiquitin enzyme complexes. Science 327:1135–1139
35. Jaattela M, Mouritzen H, Elling F, Bastholm L (1996) A20 zinc finger protein inhibits TNF and IL-1 signaling. J Immunol 156:1166–1173
36. Heyninck K, Beyaert R (1999) The cytokine-inducible zinc finger protein A20 inhibits IL-1-induced NF-kappaB activation at the level of TRAF6. FEBS Lett 442:147–150
37. Lee EG, Boone DL, Chai S, Libby SL, Chien M, Lodolce JP, Ma A (2000) Failure to regulate TNF-induced NF-κB and cell death responses in A20-deficient mice. Science 289:2350–2354
38. Reiley WW, Zhang M, Jin W, Losiewicz M, Donohue KB, Norbury CC, Sun SC (2006) Regulation of T cell development by the deubiquitinating enzyme CYLD. Nat Immunol 7:411–417
39. Reiley WW et al (2007) Deubiquitinating enzyme CYLD negatively regulates the ubiquitin-dependent kinase Tak1 and prevents abnormal T cell responses. J Exp Med 204:1475–1485
40. Carswell EA, Old LJ, Kassel RL, Green S, Fiore N, Williamson B (1975) An endotoxin-induced serum factor that causes necrosis of tumors. Proc Natl Acad Sci 72:3666–3670
41. Chen G, Goeddel DV (2002) TNF-R1 signaling: a beautiful pathway. Science 296:1634–1635
42. Tartaglia LA, Goeddel DV (1992) Two TNF receptors. Immunol Today 13:151–153
43. Micheau O, Tschopp J (2003) Induction of TNF receptor I-mediated apoptosis via two sequential signaling complexes. Cell 114:181–190
44. Wang L, Du F, Wang X (2008) TNF-alpha induces two distinct caspase-8 activation pathways. Cell 133:693–703
45. O'Donnell MA, Legarda-Addison D, Skountzos P, Yeh WC, Ting AT (2007) Ubiquitination of RIP1 regulates an NF-kappaB-independent cell-death switch in TNF signaling. Curr Biol 17:418–424
46. Mahoney DJ et al (2008) Both cIAP1 and cIAP2 regulate TNFalpha-mediated NF-kappaB activation. Proc Natl Acad Sci U S A 105:11778–11783
47. Bertrand MJ et al (2008) cIAP1 and cIAP2 facilitate cancer cell survival by functioning as E3 ligases that promote RIP1 ubiquitination. Mol Cell 30:689–700
48. Lu TT et al (2013) Dimerization and ubiquitin mediated recruitment of A20, a complex deubiquitinating enzyme. Immunity 38:896–905
49. Kelliher MA, Grimm S, Ishida Y, Kuo F, Stanger BZ, Leder P (1998) The death domain kinase RIP mediates the TNF-induced NF-kappaB signal. Immunity 8:297–303
50. Degterev A et al (2008) Identification of RIP1 kinase as a specific cellular target of necrostatins. Nat Chem Biol 4:313–321
51. Zhang H, Zhou X, McQuade T, Li J, Chan FK, Zhang J (2011) Functional complementation between FADD and RIP1 in embryos and lymphocytes. Nature 471:373–376
52. Wong WW, Gentle IE, Nachbur U, Anderton H, Vaux DL, Silke J (2010) RIPK1 is not essential for TNFR1-induced activation of NF-kappaB. Cell Death Differ 17:482–487
53. Gentle IE et al (2011) In TNF-stimulated cells, RIPK1 promotes cell survival by stabilizing TRAF2 and cIAP1, which limits induction of non-canonical NF-κB and activation of caspase-8. J Biol Chem 286:13282–13291
54. Hsu H, Shu HB, Pan MG, Goeddel DV (1996) TRADD-TRAF2 and TRADD-FADD interactions define two distinct TNF receptor 1 signal transduction pathways. Cell 84:299–308
55. Ting AT, Pimentel-Muinos FX, Seed B (1996) RIP mediates tumor necrosis factor receptor 1 activation of NF-kappaB but not Fas/APO-1-initiated apoptosis. EMBO J 15:6189–6196

56. Lee TH, Shank J, Cusson N, Kelliher MA (2004) The kinase activity of Rip1 is not required for tumor necrosis factor-alpha-induced IkappaB kinase or p38 MAP kinase activation or for the ubiquitination of Rip1 by Traf2. J Biol Chem 279:33185–33191
57. Zhang SQ, Kovalenko A, Cantarella G, Wallach D (2000) Recruitment of the IKK signalosome to the p55 TNF receptor: RIP and A20 bind to NEMO (IKKgamma) upon receptor stimulation. Immunity 12:301–311
58. Devin A, Cook A, Lin Y, Rodriguez Y, Kelliher M, Liu Z (2000) The distinct roles of TRAF2 and RIP in IKK activation by TNF-R1: TRAF2 recruits IKK to TNF-R1 while RIP mediates IKK activation. Immunity 12:419–429
59. Ea CK, Deng L, Xia ZP, Pineda G, Chen ZJ (2006) Activation of IKK by TNFalpha requires site-specific ubiquitination of RIP1 and polyubiquitin binding by NEMO. Mol Cell 22:245–257
60. Newton K et al (2008) Ubiquitin chain editing revealed by polyubiquitin linkage-specific antibodies. Cell 134:668–678
61. Yeh WC et al (1997) Early lethality, functional NF-kappaB activation, and increased sensitivity to TNF-induced cell death in TRAF2-deficient mice. Immunity 7:715–725
62. Tada K et al (2001) Critical roles of TRAF2 and TRAF5 in tumor necrosis factor-induced NF-κB activation and protection from cell death. J Biol Chem 276:36530–36534
63. Wertz IE et al (2004) De-ubiquitination and ubiquitin ligase domains of A20 downregulate NF-kappaB signalling. Nature 430:694–699
64. Yin Q, Lamothe B, Darnay BG, Wu H (2009) Structural basis for the lack of E2 interaction in the RING domain of TRAF2. Biochemistry 48:10558–10567
65. Fukushima T, S-i M, Kress CL, Bruey JM, Krajewska M, Lefebvre S, Zapata JM, Ze R, Reed JC (2007) Ubiquitin-conjugating enzyme Ubc13 is a critical component of TNF receptor-associated factor (TRAF)-mediated inflammatory responses. Proc Natl Acad Sci 104:6371–6376
66. Kirisako T et al (2006) A ubiquitin ligase complex assembles linear polyubiquitin chains. EMBO J 25:4877–4887
67. Haas TL et al (2009) Recruitment of the linear ubiquitin chain assembly complex stabilizes the TNF-R1 signaling complex and is required for TNF-mediated gene induction. Mol Cell 36:831–844
68. Tokunaga F et al (2009) Involvement of linear polyubiquitylation of NEMO in NF-kappaB activation. Nat Cell Biol 11:123–132
69. Seymour RE, Hasham MG, Cox GA, Shultz LD, Hogenesch H, Roopenian DC, Sundberg JP (2007) Spontaneous mutations in the mouse Sharpin gene result in multiorgan inflammation, immune system dysregulation and dermatitis. Genes Immun 8:416–421
70. Gerlach B et al (2011) Linear ubiquitination prevents inflammation and regulates immune signalling. Nature 471:591–596
71. Ikeda F et al (2011) SHARPIN forms a linear ubiquitin ligase complex regulating NF-kappaB activity and apoptosis. Nature 471:637–641
72. Zak DE et al (2011) Systems analysis identifies an essential role for SHANK-associated RH domain-interacting protein (SHARPIN) in macrophage Toll-like receptor 2 (TLR2) responses. Proc Natl Acad Sci U S A 108:11536–11541
73. Tokunaga F, Nakagawa T, Nakahara M, Saeki Y, Taniguchi M, Sakata S, Tanaka K, Nakano H, Iwai K (2011) SHARPIN is a component of the NF-kappaB-activating linear ubiquitin chain assembly complex. Nature 471:633–636
74. Liang Y, Seymour RE, Sundberg JP (2011) Inhibition of NF-kappaB signaling retards eosinophilic dermatitis in SHARPIN-deficient mice. J Invest Dermatol 131:141–149
75. Rantala JK et al (2011) SHARPIN is an endogenous inhibitor of beta1-integrin activation. Nat Cell Biol 13:1315–1324
76. Enesa K, Zakkar M, Chaudhury H, Luong LA, Rawlinson L, Mason JC, Haskard DO, Dean JLE, Evans PC (2008) NF-κB suppression by the deubiquitinating enzyme cezanne: a novel negative feedback loop in pro-inflammatory signaling. J Biol Chem 283:7036–7045
77. Nijman SM, Luna-Vargas MP, Velds A, Brummelkamp TR, Dirac AM, Sixma TK, Bernards R (2005) A genomic and functional inventory of deubiquitinating enzymes. Cell 123:773–786

78. Amerik AY, Hochstrasser M (2004) Mechanism and function of deubiquitinating enzymes. Biochim Biophys Acta 1695:189–207
79. Harhaj EW, Dixit VM (2011) Deubiquitinases in the regulation of NF-[kappa]B signaling. Cell Res 21:22–39
80. Reiley W, Zhang M, Wu X, Granger E, Sun SC (2005) Regulation of the deubiquitinating enzyme CYLD by IkappaB kinase gamma-dependent phosphorylation. Mol Cell Biol 25:3886–3895
81. Sun SC (2008) Deubiquitylation and regulation of the immune response. Nat Rev Immunol 8:501–511
82. Hutti JE, Turk BE, Asara JM, Ma A, Cantley LC, Abbott DW (2007) IkappaB kinase beta phosphorylates the K63 deubiquitinase A20 to cause feedback inhibition of the NF-kappaB pathway. Mol Cell Biol 27:7451–7461
83. Komander D, Lord CJ, Scheel H, Swift S, Hofmann K, Ashworth A, Barford D (2008) The structure of the CYLD USP domain explains its specificity for Lys63-linked polyubiquitin and reveals a B box module. Mol Cell 29:451–464
84. Jin W et al (2008) Deubiquitinating enzyme CYLD negatively regulates RANK signaling and osteoclastogenesis in mice. J Clin Invest 118:1858–1866
85. Komander D, Reyes-Turcu F, Licchesi JD, Odenwaelder P, Wilkinson KD, Barford D (2009) Molecular discrimination of structurally equivalent Lys 63-linked and linear polyubiquitin chains. EMBO Rep 10:466–473
86. Lin SC, Chung JY, Lamothe B, Rajashankar K, Lu M, Lo YC, Lam AY, Darnay BG, Wu H (2008) Molecular basis for the unique deubiquitinating activity of the NF-kappaB inhibitor A20. J Mol Biol 376:526–540
87. De Valck D, Jin DY, Heyninck K, Van de Craen M, Contreras R, Fiers W, Jeang KT, Beyaert R (1999) The zinc finger protein A20 interacts with a novel anti-apoptotic protein which is cleaved by specific caspases. Oncogene 18:4182–4190
88. Opipari AW Jr, Boguski MS, Dixit VM (1990) The A20 cDNA induced by tumor necrosis factor alpha encodes a novel type of zinc finger protein. J Biol Chem 265:14705–14708
89. Opipari AW Jr, Hu HM, Yabkowitz R, Dixit VM (1992) The A20 zinc finger protein protects cells from tumor necrosis factor cytotoxicity. J Biol Chem 267:12424–12427
90. Tewari M, Wolf FW, Seldin MF, O'Shea KS, Dixit VM, Turka LA (1995) Lymphoid expression and regulation of A20, an inhibitor of programmed cell death. J Immunol 154:1699–1706
91. Evans PC, Ovaa H, Hamon M, Kilshaw PJ, Hamm S, Bauer S, Ploegh HL, Smith TS (2004) Zinc-finger protein A20, a regulator of inflammation and cell survival, has de-ubiquitinating activity. Biochem J 378:727–734
92. Song HY, Rothe M, Goeddel DV (1996) The tumor necrosis factor-inducible zinc finger protein A20 interacts with TRAF1/TRAF2 and inhibits NF-kappaB activation. Proc Natl Acad Sci U S A 93:6721–6725
93. Heyninck K, De Valck D, Vanden Berghe W, Van Criekinge W, Contreras R, Fiers W, Haegeman G, Beyaert R (1999) The zinc finger protein A20 inhibits TNF-induced NF-kappaB-dependent gene expression by interfering with an RIP- or TRAF2-mediated transactivation signal and directly binds to a novel NF-kappaB-inhibiting protein ABIN. J Cell Biol 145:1471–1482
94. Li L, Hailey DW, Soetandyo N, Li W, Lippincott-Schwartz J, Shu HB, Ye Y (2008) Localization of A20 to a lysosome-associated compartment and its role in NFkappaB signaling. Biochim Biophys Acta 1783:1140–1149
95. Li L, Soetandyo N, Wang Q, Ye Y (2009) The zinc finger protein A20 targets TRAF2 to the lysosomes for degradation. Biochim Biophys Acta 1793:346–353
96. Skaug B, Chen J, Du F, He J, Ma A, Chen ZJ (2011) Direct, noncatalytic mechanism of IKK inhibition by A20. Mol Cell 44:559–571
97. Boone DL et al (2004) The ubiquitin-modifying enzyme A20 is required for termination of Toll-like receptor responses. Nat Immunol 5:1052–1060
98. Turer EE, Tavares RM, Mortier E, Hitotsumatsu O, Advincula R, Lee B, Shifrin N, Malynn BA, Ma A (2008) Homeostatic MyD88-dependent signals cause lethal inflammation in the absence of A20. J Exp Med 205:451–464

99. Vereecke L, Sze M, Mc Guire C, Rogiers B, Chu Y, Schmidt-Supprian M, Pasparakis M, Beyaert R, van Loo G (2010) Enterocyte-specific A20 deficiency sensitizes to tumor necrosis factor-induced toxicity and experimental colitis. J Exp Med 207:1513–1523
100. Matmati M et al (2011) A20 (TNFAIP3) deficiency in myeloid cells triggers erosive polyarthritis resembling rheumatoid arthritis. Nat Genet 43:908–912
101. Song XT, Evel-Kabler K, Shen L, Rollins L, Huang XF, Chen SY (2008) A20 is an antigen presentation attenuator, and its inhibition overcomes regulatory T cell-mediated suppression. Nat Med 14:258–265
102. Kool M et al (2011) The ubiquitin-editing protein A20 prevents dendritic cell activation, recognition of apoptotic cells, and systemic autoimmunity. Immunity 35:82–96
103. Graham RR et al (2008) Genetic variants near TNFAIP3 on 6q23 are associated with systemic lupus erythematosus. Nat Genet 40:1059–1061
104. Stilo R, Varricchio E, Liguoro D, Leonardi A, Vito P (2008) A20 is a negative regulator of BCL10- and CARMA3-mediated activation of NF-kappaB. J Cell Sci 121:1165–1171
105. Duwel M et al (2009) A20 negatively regulates T cell receptor signaling to NF-kappaB by cleaving Malt1 ubiquitin chains. J Immunol 182:7718–7728
106. Coornaert B et al (2008) T cell antigen receptor stimulation induces MALT1 paracaspase-mediated cleavage of the NF-kappaB inhibitor A20. Nat Immunol 9:263–271
107. Sarma V, Lin Z, Clark L, Rust BM, Tewari M, Noelle RJ, Dixit VM (1995) Activation of the B-cell surface receptor CD40 induces A20, a novel zinc finger protein that inhibits apoptosis. J Biol Chem 270:12343–12346
108. Tavares RM et al (2010) The ubiquitin modifying enzyme A20 restricts B cell survival and prevents autoimmunity. Immunity 33:181–191
109. Ferch U, Kloo B, Gewies A, Pfander V, Duwel M, Peschel C, Krappmann D, Ruland J (2009) Inhibition of MALT1 protease activity is selectively toxic for activated B cell-like diffuse large B cell lymphoma cells. J Exp Med 206:2313–2320
110. Iha H et al (2008) Inflammatory cardiac valvulitis in TAX1BP1-deficient mice through selective NF-kappaB activation. EMBO J 27:629–641
111. Shembade N, Harhaj NS, Parvatiyar K, Copeland NG, Jenkins NA, Matesic LE, Harhaj EW (2008) The E3 ligase Itch negatively regulates inflammatory signaling pathways by controlling the function of the ubiquitin-editing enzyme A20. Nat Immunol 9:254–262
112. Wagner S et al (2008) Ubiquitin binding mediates the NF-kappaB inhibitory potential of ABIN proteins. Oncogene 27:3739–3745
113. Mauro C, Pacifico F, Lavorgna A, Mellone S, Iannetti A, Acquaviva R, Formisano S, Vito P, Leonardi A (2006) ABIN-1 binds to NEMO/IKKgamma and co-operates with A20 in inhibiting NF-kappaB. J Biol Chem 281:18482–18488
114. Gachon F, Peleraux A, Thebault S, Dick J, Lemasson I, Devaux C, Mesnard JM (1998) CREB-2, a cellular CRE-dependent transcription repressor, functions in association with Tax as an activator of the human T-cell leukemia virus type 1 promoter. J Virol 72:8332–8337
115. Shembade N, Harhaj NS, Liebl DJ, Harhaj EW (2007) Essential role for TAX1BP1 in the termination of TNF-alpha-, IL-1- and LPS-mediated NF-kappaB and JNK signaling. EMBO J 26:3910–3922
116. Sudol M, Chen HI, Bougeret C, Einbond A, Bork P (1995) Characterization of a novel protein-binding module–the WW domain. FEBS Lett 369:67–71
117. Perry WL, Hustad CM, Swing DA, O'Sullivan TN, Jenkins NA, Copeland NG (1998) The itchy locus encodes a novel ubiquitin protein ligase that is disrupted in a18H mice. Nat Genet 18:143–146
118. Colland F et al (2004) Functional proteomics mapping of a human signaling pathway. Genome Res 14:1324–1332
119. Subramaniam V, Li H, Wong M, Kitching R, Attisano L, Wrana J, Zubovits J, Burger AM, Seth A (2003) The RING-H2 protein RNF11 is overexpressed in breast cancer and is a target of Smurf2 E3 ligase. Br J Cancer 89:1538–1544
120. Kitching R, Wong MJ, Koehler D, Burger AM, Landberg G, Gish G, Seth A (2003) The RING-H2 protein RNF11 is differentially expressed in breast tumours and interacts with HECT-type E3 ligases. Biochim Biophys Acta 1639:104–112

121. Shembade N, Parvatiyar K, Harhaj NS, Harhaj EW (2009) The ubiquitin-editing enzyme A20 requires RNF11 to downregulate NF-kappaB signalling. EMBO J 28:513–522
122. Azmi P, Seth A (2005) RNF11 is a multifunctional modulator of growth factor receptor signalling and transcriptional regulation. Eur J Cancer 41:2549–2560
123. Li H, Seth A (2004) An RNF11: Smurf2 complex mediates ubiquitination of the AMSH protein. Oncogene 23:1801–1808
124. Heyninck K, Kreike MM, Beyaert R (2003) Structure-function analysis of the A20-binding inhibitor of NF-kappa B activation, ABIN-1. FEBS Lett 536:135–140
125. Ashida H, Kim M, Schmidt-Supprian M, Ma A, Ogawa M, Sasakawa C (2010) A bacterial E3 ubiquitin ligase IpaH9.8 targets NEMO/IKKgamma to dampen the host NF-kappaB-mediated inflammatory response. Nat Cell Biol 12(1):66–73, sup pp 61–69
126. Oshima S et al (2009) ABIN-1 is a ubiquitin sensor that restricts cell death and sustains embryonic development. Nature 457:906–909
127. Bohgaki M, Tsukiyama T, Nakajima A, Maruyama S, Watanabe M, Koike T, Hatakeyama S (2008) Involvement of Ymer in suppression of NF-kappaB activation by regulated interaction with lysine-63-linked polyubiquitin chain. Biochim Biophys Acta 1783:826–837
128. Fenner BJ, Scannell M, Prehn JH (2009) Identification of polyubiquitin binding proteins involved in NF-kappaB signaling using protein arrays. Biochim Biophys Acta 1794:1010–1016
129. Fung EY, Smyth DJ, Howson JM, Cooper JD, Walker NM, Stevens H, Wicker LS, Todd JA (2009) Analysis of 17 autoimmune disease-associated variants in type 1 diabetes identifies 6q23/TNFAIP3 as a susceptibility locus. Genes Immun 10:188–191
130. Compagno M et al (2009) Mutations of multiple genes cause deregulation of NF-kappaB in diffuse large B-cell lymphoma. Nature 459:717–721
131. Kato M et al (2009) Frequent inactivation of A20 in B-cell lymphomas. Nature 459:712–716
132. Musone SL et al (2008) Multiple polymorphisms in the TNFAIP3 region are independently associated with systemic lupus erythematosus. Nat Genet 40:1062–1064
133. Schmitz R et al (2009) TNFAIP3 (A20) is a tumor suppressor gene in Hodgkin lymphoma and primary mediastinal B cell lymphoma. J Exp Med 206:981–989
134. Vereecke L, Beyaert R, van Loo G (2009) The ubiquitin-editing enzyme A20 (TNFAIP3) is a central regulator of immunopathology. Trends Immunol 30:383–391
135. Trynka G et al (2009) Coeliac disease-associated risk variants in TNFAIP3 and REL implicate altered NF-kappaB signalling. Gut 58:1078–1083
136. Wang K et al (2010) Comparative genetic analysis of inflammatory bowel disease and type 1 diabetes implicates multiple loci with opposite effects. Hum Mol Genet 19:2059–2067
137. Boonyasrisawat W et al (2007) Tag polymorphisms at the A20 (TNFAIP3) locus are associated with lower gene expression and increased risk of coronary artery disease in type 2 diabetes. Diabetes 56:499–505
138. Nair RP et al (2009) Genome-wide scan reveals association of psoriasis with IL-23 and NF-kappaB pathways. Nat Genet 41:199–204
139. Lodolce JP et al (2010) African-derived genetic polymorphisms in TNFAIP3 mediate risk for autoimmunity. J Immunol 184:7001–7009
140. Thelander EF et al (2008) Characterization of 6q deletions in mature B cell lymphomas and childhood acute lymphoblastic leukemia. Leuk Lymphoma 49:477–487
141. Honma K, Tsuzuki S, Nakagawa M, Tagawa H, Nakamura S, Morishima Y, Seto M (2009) TNFAIP3/A20 functions as a novel tumor suppressor gene in several subtypes of non-Hodgkin lymphomas. Blood 114:2467–2475
142. Chanudet E et al (2010) A20 is targeted by promoter methylation, deletion and inactivating mutation in MALT lymphoma. Leukemia 24:483–487
143. Huang HL, Yeh WC, Lai MZ, Mirtsos C, Chau H, Chou CH, Benchimol S (2009) Impaired TNFalpha-induced A20 expression in E1A/Ras-transformed cells. Br J Cancer 101:1555–1564
144. Vendrell JA, Ghayad S, Ben-Larbi S, Dumontet C, Mechti N, Cohen PA (2007) A20/TNFAIP3, a new estrogen-regulated gene that confers tamoxifen resistance in breast cancer cells. Oncogene 26:4656–4667

145. Hjelmeland AB et al (2010) Targeting A20 decreases glioma stem cell survival and tumor growth. PLoS Biol 8:e1000319
146. Harhaj EW, Harhaj NS (2005) Mechanisms of persistent NF-kappaB activation by HTLV-I tax. IUBMB Life 57:83–91
147. Chin KT, Chun AC, Ching YP, Jeang KT, Jin DY (2007) Human T-cell leukemia virus oncoprotein tax represses nuclear receptor-dependent transcription by targeting coactivator TAX1BP1. Cancer Res 67:1072–1081
148. Wang X, Naidu SR, Sverdrup F, Androphy EJ (2009) Tax1BP1 interacts with papillomavirus E2 and regulates E2-dependent transcription and stability. J Virol 83:2274–2284
149. Laherty CD, Hu HM, Opipari AW, Wang F, Dixit VM (1992) The Epstein-Barr virus LMP1 gene product induces A20 zinc finger protein expression by activating nuclear factor kappa B. J Biol Chem 267:24157–24160
150. Fries KL, Miller WE, Raab-Traub N (1999) The A20 protein interacts with the Epstein-Barr virus latent membrane protein 1 (LMP1) and alters the LMP1/TRAF1/TRADD complex. Virology 264:159–166
151. Ning S, Pagano JS (2010) The A20 deubiquitinase activity negatively regulates LMP1 activation of IRF7. J Virol 84:6130–6138
152. Whitehouse CA (2004) Crimean-Congo hemorrhagic fever. Antiviral Res 64:145–160
153. Zhou H, Monack DM, Kayagaki N, Wertz I, Yin J, Wolf B, Dixit VM (2005) Yersinia virulence factor YopJ acts as a deubiquitinase to inhibit NF-kappa B activation. J Exp Med 202:1327–1332
154. Komander D, Barford D (2008) Structure of the A20 OTU domain and mechanistic insights into deubiquitination. Biochem J 409:77–85
155. Wang YY, Li L, Han KJ, Zhai Z, Shu HB (2004) A20 is a potent inhibitor of TLR3- and Sendai virus-induced activation of NF-kappaB and ISRE and IFN-beta promoter. FEBS Lett 576:86–90
156. Lin R, Yang L, Nakhaei P, Sun Q, Sharif-Askari E, Julkunen I, Hiscott J (2006) Negative regulation of the retinoic acid-inducible gene I-induced antiviral state by the ubiquitin-editing protein A20. J Biol Chem 281:2095–2103
157. Parvatiyar K, Barber GN, Harhaj EW (2010) TAX1BP1 and A20 inhibit antiviral signaling by targeting TBK1-IKKi kinases. J Biol Chem 285:14999–15009

Chapter 3
The Deubiquitinase Activity of A20 Is Dispensable for Its Role in NF-κB Signaling

3.1 Introduction

NF-κB is an ubiquitously expressed, inducible transcription factor that regulates the expression of numerous target genes, particularly in the immune system. In unstimulated cells, NF-κB is sequestered in the cytoplasm through its binding to inhibitory IκB proteins that mask the nuclear localizing signals on the NF-κB proteins. NF-κB can be activated upon stimulation of cells with a wide variety of inducers including proinflammatory cytokines such as TNFα and IL-1β, or components of pathogenic microbes (PAMPs) such as bacterial lipopolysaccharide (LPS). Upon engagement of cognate receptors e.g., TNF receptor or Toll-like receptor 4, signaling pathways are triggered that lead to the activation of a protein kinase complex known as the IκB kinase (IKK). The activated IKK phosphorylates the IκB proteins on specific serine residues that lead to their polyubiquitination with K48-linked ubiquitin chains, and consequent degradation by the proteasome. The released NF-κB then migrates to the nucleus where it binds to promoters of target genes and activates transcription [1].

Besides the well-characterized K48-linked polyubiquitination that triggers proteasome-mediated degradation of substrate proteins such as IκBs, polyubiquitination of signaling adapter proteins with K63-linked chains have been proposed to have critical regulatory function in NF-κB activation pathways. RING-finger proteins such as TRAFs are believed to act as ubiquitinating enzymes and molecules such as RIP1 are well-characterized substrates that undergo K63-linked ubiquitination. Similar to kinases and phosphatases, deubiquitinating enzymes have been proposed to act to limit the consequence of regulatory ubiquitination by removing the

---

This chapter was originally published as *The deubiquitinase activity of A20 is dispensable for NF-κB signaling* in *EMBO Reports*, John Wiley & Sons, Ltd, 15(7), 2014, 775–783, http://dx.doi.org/10.15252/embr.201338305. First published online 30 MAY 2014 ©Arnab De, Teruki Dainichi, Chozha Vendan Rathinam and Sankar Ghosh.

A. De, *Ubiquitin Chains: Degradation and Beyond*, Springer Theses,
DOI 10.1007/978-3-319-14965-3_3

polyubiquitin chains. Amongst the various deubiquitinases that have been suggested to act on K63-linked chains in the NF-κB pathway, the best characterized is A20 (TNFAIP3) [2].

A20 was initially identified as a TNFα-inducible zinc-finger protein that protects cells from TNFα-induced cytotoxicity [3]. A20 is expressed at very low levels in most cell types but is rapidly induced in response to various PAMPs or proinflammatory cytokines [3]. Subsequently, many studies, mainly in overexpression systems, have demonstrated that A20 downregulates NF-κB signaling in multiple pathways including the tumor necrosis factor (TNFα) and Toll-like receptor (TLR) pathways [4–6]. Mice deficient in A20 die prematurely from multiorgan inflammation and cachexia as a result of increased NF-κB signaling as would be predicted from the in vitro studies. These mice are also sensitive to proinflammatory stimuli and cannot withstand even sub-lethal doses of TNFα or LPS. Additionally, the A20-deficient MEFs show persistent NF-κB signaling as evidenced by increased IκBα degradation [7].

Intriguingly, although A20 was initially identified as a TNFα-inducible gene, the spontaneous inflammation in A20-deficient mice continues unabated even in A20(TNFAIP3)$^{-/-}$ TNFα$^{-/-}$ and A20(TNFAIP3)$^{-/-}$ TNFR1$^{-/-}$ double mutant mice. Thus, the spontaneous inflammation in A20-deficient mice results from TNFα-independent signaling. Instead the inflammation in these mice is largely eliminated by removal of MyD88, and the TNFAIP3$^{-/-}$ Myd88$^{-/-}$ mice do not show the severe inflammation characteristic of A20-deficient mice, indicating that TLR signaling drives the spontaneous inflammation in these mice [8]. Consistently, treatment of A20-deficient mice with broad-spectrum antibiotics rescues the inflammatory phenotype, thereby indicating that a dysregulated intestinal flora contributes to the constitutive TLR signaling which results in perinatal lethality in these mice [9]. Thus, it has been clearly demonstrated that A20 is essential and nonredundant in restricting persistent TLR-mediated activation of NF-κB and subsequent lethality. Therefore while many studies have focused on the role of A20 in response to TNFα, it is imperative to understand the role of A20 in the context of TLR and LPS signaling [10, 11].

It has been shown by various biochemical assays that A20 can function both as a deubiquitinase (DUB) and as an ubiquitin ligase. A20 has a N-terminal OTU domain followed by seven C-terminal zinc-finger domains. Cys-103 in the OTU domain has been showed to be essential for the deubiquitinating function of A20 [11, 12]. Previous biochemical studies using recombinant A20 has demonstrated that A20 is a unique ubiquitin modifying enzyme regulating both the activity (by removing K63-linked chains) and stability (by adding K48 linked chains) of signaling molecules such as RIP and TRAF6 (Nature, 2004, 430, 694–699) [11]. These studies showed that in response to TNFα stimulation, A20 acts by first deubiquitinating the regulatory K63-linked polyubiquitin chains from RIP1, thereby initially attenuating NF-κB signaling. Subsequently, A20 acts as an E3 ubiquitin ligase adding K48-linked polyubiquitin chains to RIP1 leading to degradation of RIP1 and terminating any residual NF-κB signaling. In the TLR and the IL-1R pathway, A20 inhibits NF-κB signaling by disrupting the binding of the E3 ligase, TRAF6, with E2 ubiquitin conjugating enzymes like Ubc13 or UbcH5c. The deubiquitinase activity mediated by C103 residue has been shown to be essential in mediating this interaction as well [13].

Despite the A20 C103A mutants being completely deficient in DUB activity, some studies have shown that overexpression of A20 C103A can also inhibit NF-κB signaling [4, 12, 14]. We were intrigued by these observations because they suggested that the ability of A20 to inhibit NF-κB activation might not be due to the ability of A20 to deubiquitinate K63-ubiquitinated substrates. Therefore, to accurately delineate the physiological ramifications of A20's deubiquitinating function, we generated a gene-targeted mouse with a C103A point mutation to eliminate the DUB activity of A20. Given the severe multiorgan inflammation in $A20^{-/-}$ mice that led to perinatal lethality [7], we hypothesized that these A20-C103A knock-in deubiquitinase mutant mice would show a phenotype similar to the A20 knock-out mice. Furthermore since dysregulation of A20 has been implicated in a host of diseases [15], these knock-in mice could prove to be a potential model for studying these disorders in a physiologically relevant setting. However, as described in this report, the A20 C103A KI mice do not show any aberrant proinflammatory mice and demonstrate normal lifespan without any observable phenotype. Analysis of NF-κB signaling in cells isolated from these mice show no discernible difference from normal, wild-type cells, including the degree of ubiquitination of signaling adapter proteins. Therefore our studies reveal that the deubiquitinase function of A20 is not important for the well-described role of A20 in NF-κB signaling.

3.2 Results

3.2.1 Generation of A20-OTU Knock-In Mice

In order to evaluate the physiological contribution of A20's deubiquitinating function, we decided to create a mouse with a mutation in the OTU domain to abolish the deubiquitinase activity. We used a BAC-based approach, as described in the methods section, to generate a knock-in mouse that harbors a point mutation at position 103, namely a change from cysteine to alanine (Fig. 3.1A). Heterozygous A20 knock-in mice (C103A or C/A) were intercrossed to generate the homozygous A20-OTU knock-in mice (or A/A mice). Sequencing of genomic DNA from both C/A and A/A mice confirmed the presence of the correct C103A point mutation (Fig. 3.1B). The sequencing chromatogram of the genomic DNA isolated from heterozygous mice shows both cysteine and alanine at position 103, while the homozygous mice encodes for only alanine. The genotyping strategy and the photographs of the respective littermates are shown in Fig. 3.1C. In this manuscript, we refer to the wild-type littermates as C/C meaning they have a cysteine in both alleles at position 103; the heterozygous mice as C/A as they have a cysteine and alanine in the two alleles; and the homozygous knock-in mice as A/A as they have only the mutated alanines on both alleles.

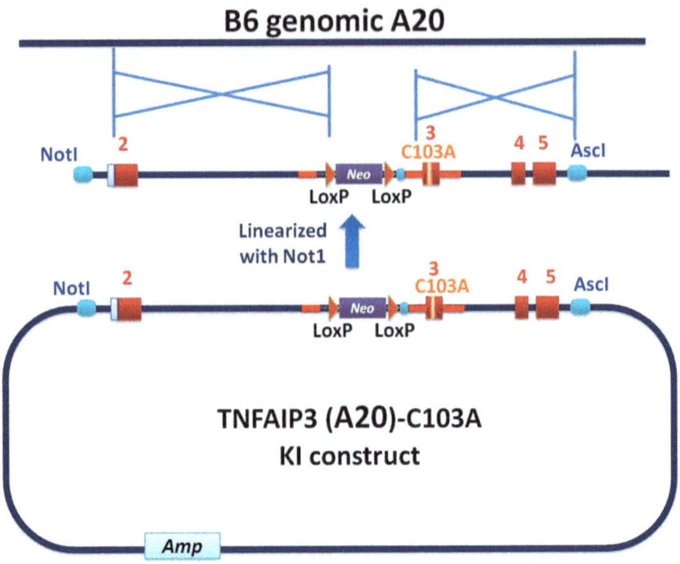

Fig. 3.1A Gene-targeting strategy for generating the A20OTU/OTU mice: The plasmid construct was linearized using Not1 and chimeras were obtained after homologous recombination with the B6-genomic A20 in the ES cells (Copyright: Arnab De, Teruki Dainichi, Chozha Vendan Rathinam and Sankar Ghosh; previously published in EMBO Reports, 15(7), 2014, 775-783)

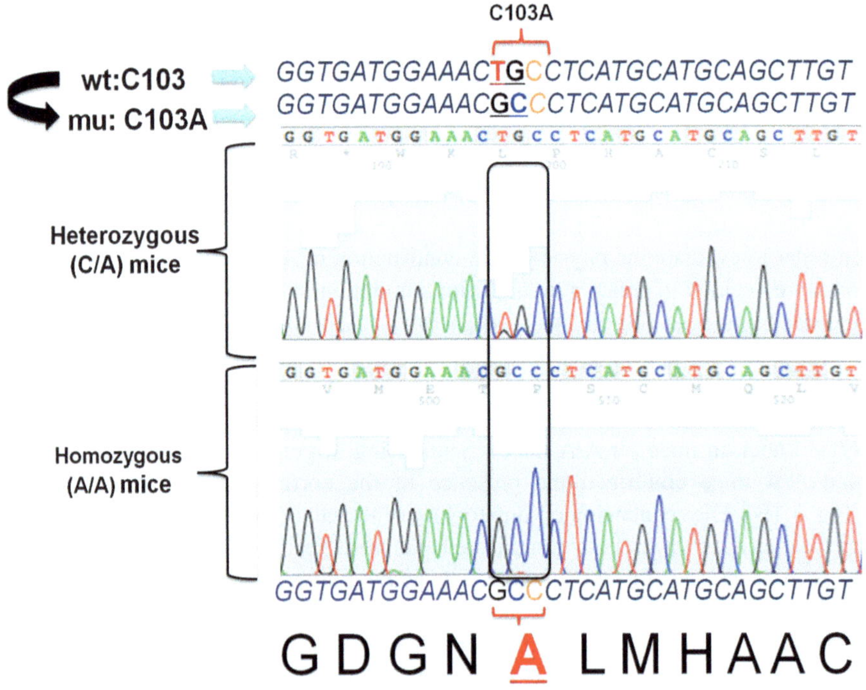

Fig. 3.1B Sequencing of genomic DNA from homozygous (A/A) and heterozygous littermates (C/A) (Copyright: Arnab De, Teruki Dainichi, Chozha Vendan Rathinam and Sankar Ghosh; previously published in EMBO Reports, 15(7), 2014, 775-783)

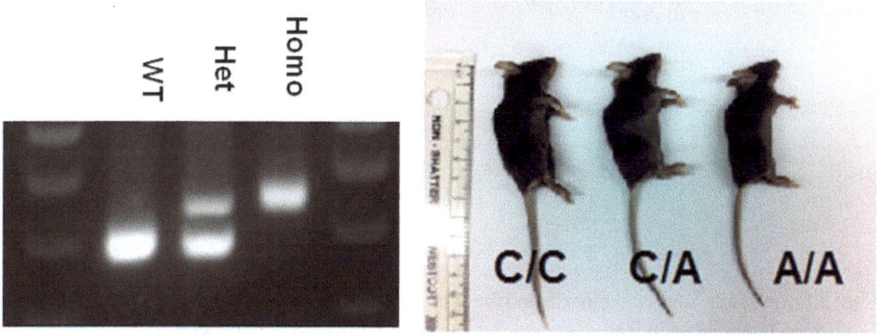

Fig. 3.1C Genotyping strategy and photograph of wild-type, heterozygous and homozygous littermates (Copyright: Arnab De, Teruki Dainichi, Chozha Vendan Rathinam and Sankar Ghosh; previously published in EMBO Reports, 15(7), 2014, 775-783)

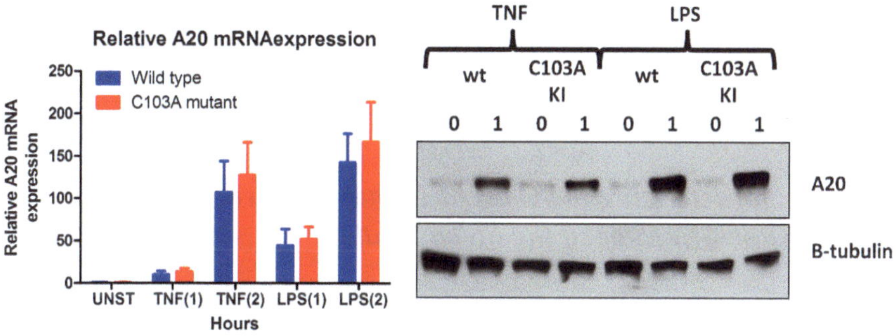

Fig. 3.1D Relative A20 mRNA and protein expression in macrophages from littermates. Error bars represent standard deviation. Data represents 3 mice for each genotype (Copyright: Arnab De, Teruki Dainichi, Chozha Vendan Rathinam and Sankar Ghosh; previously published in EMBO Reports, 15(7), 2014, 775-783)

We next isolated BMDMs from both WT and A/A littermates and found roughly equal levels of A20 mRNA and protein (Fig. 3.1D) following LPS and TNFα stimulation. A20 was initially found at relatively low levels but is rapidly induced upon LPS stimulation in both the C/C and A/A littermates.

3.2.2 *Characterization of A20 OTU Knock-In Mice*

Although A20$^{-/-}$ mice were runted by 1 week of age and die perinatally of multiorgan inflammation and cachexia, the homozygous A/A knock-in mice were normal (Fig. 3.1E) and did not display any external signs of inflammation (for the observed period of 10 months after birth). They were born in Mendelian ratios and both the C/C and A/A littermate adults weighed the same (Fig. 3.1E). Histological examination

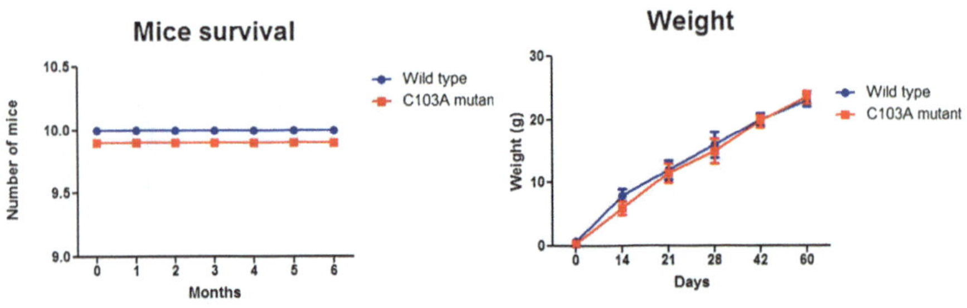

Fig. 3.1E Normal survival curves and weight of littermates (Copyright: Arnab De, Teruki Dainichi, Chozha Vendan Rathinam and Sankar Ghosh; previously published in EMBO Reports, 15(7), 2014, 775-783)

of 6-week-old A20$^{-/-}$ mice had revealed severe tissue damage in multiple organs. However histology of major organs (kidney, liver, spleen, thymus, lung, and heart) of WT (C/C), heterozygous knock-in (C/A) and homozygous knock-in (A/A) littermates did not reveal any differences (Fig. 3.2A). Hence, the deubiquitinating function of A20 is not important for maintaining basal tissue homeostasis.

3.2.3 Characterizing the Cells of Immune System of C103A Knock-In Mice in the Steady State

Unlike A20$^{-/-}$ mice that have increased numbers of myeloid lineage cells in the bone marrow and spleen, the C/C, C/A, and A/A littermates had comparable frequencies of myeloid lineage cells (CD11b+). Within the myeloid compartment, the frequencies of monocytes (CD11b+Ly6c+Ly6G−) and granulocytes (CD11b+Ly6c+Ly6G+) were similar across all indicated genotypes (Fig. 3.2B). Similarly, analysis of B-cell development indicated that the differentiation pattern of CD19+B220+ and CD19−B220+ cells in heterozygous and homozygous knock-in mice was comparable to the WT littermates. In addition, analysis of surface IgM and IgD expression in CD19+ B cells of the bone marrow indicated similar expression patterns amongst wild-type, heterozygous and homozygous knock-in mice (Fig. 3.2C). Consistently, analysis of erythroid (Ter119+) and megakaryocyte (CD41+) lineage cells suggested comparable frequencies in the bone marrow (Fig. 3.2D).

In mice, T-cell development occurs in the thymus and undergoes distinct stages of differentiation including the, earliest, double negative (DN) stage (CD4−CD8−), double positive (DP) stage (CD4+CD8+), and subsequently into CD4+CD8− and CD4−CD8+ single positive cells [16]. Analysis of thymus of the three genotypes indicated a roughly equal differentiation of CD4 and CD8 cells between the genotypes as shown in Fig. 3.2E. Consistent with the normal differentiation of myeloid, lymphoid, and erythroid lineages in the BM and thymus, the spleen also showed normal proportions of B cells (Fig. 3.2F), T cells (Fig. 3.2G), dendritic cells

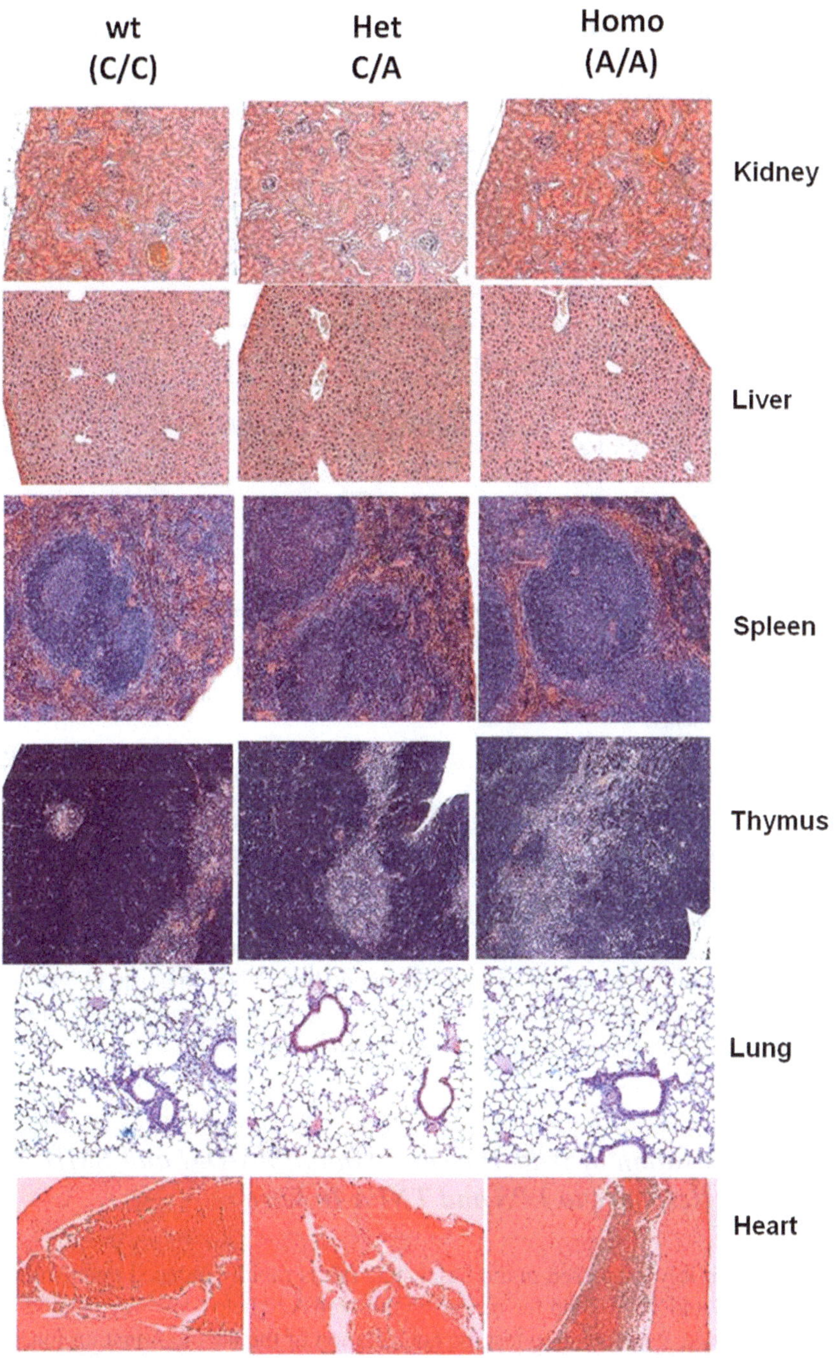

Fig. 3.2A Histology of kidney, liver, spleen, thymus, lung, and heart of littermates (Copyright: Arnab De, Teruki Dainichi, Chozha Vendan Rathinam and Sankar Ghosh; previously published in EMBO Reports, 15(7), 2014, 775-783)

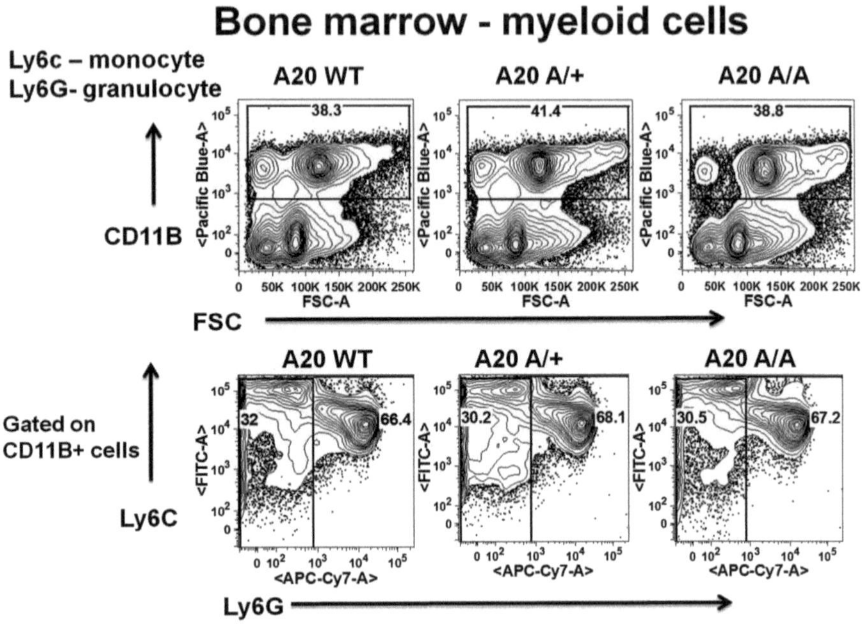

Fig. 3.2B Characterizing the myeloid cells of the bone marrow (Copyright: Arnab De, Teruki Dainichi, Chozha Vendan Rathinam and Sankar Ghosh; previously published in EMBO Reports, 15(7), 2014, 775-783)

(Fig. 3.2H), and myeloid cells (Fig. 3.2I). Therefore these results suggest that the deubiquitinase function of A20 does not play a role in differentiation and maintenance of immune cell types.

As mice age to 6 months, the A/A deubiquitinase mutant mice develop splenomegaly and had larger spleens as compared to their C/C littermates. Both the wild-type and mutant mice weigh around the same (Fig. 3.2J). Analysis of the bone marrow and spleen of older mice showed that the homozygous mutant mice showed an increase in the numbers of myeloid cells (Fig. 3.2K).

3.2.4 Activation of Wild-Type and Mutant BMDM and BMDCs in Response to LPS and TNFα In Vitro

A20 is negligibly expressed in the basal state but is rapidly induced upon stimulation by LPS and TNFα, and the newly synthesized A20 helps downregulate NF-κB activity. Hence, we wanted to test the contribution of the deubiquitinating function of A20 in inhibiting NF-κB activation following stimulation with LPS and TNFα.

We differentiated bone marrow cells of wild-type (C/C), heterozygous (C/A), and homozygous (A/A) mice into dendritic cells and macrophages in the presence

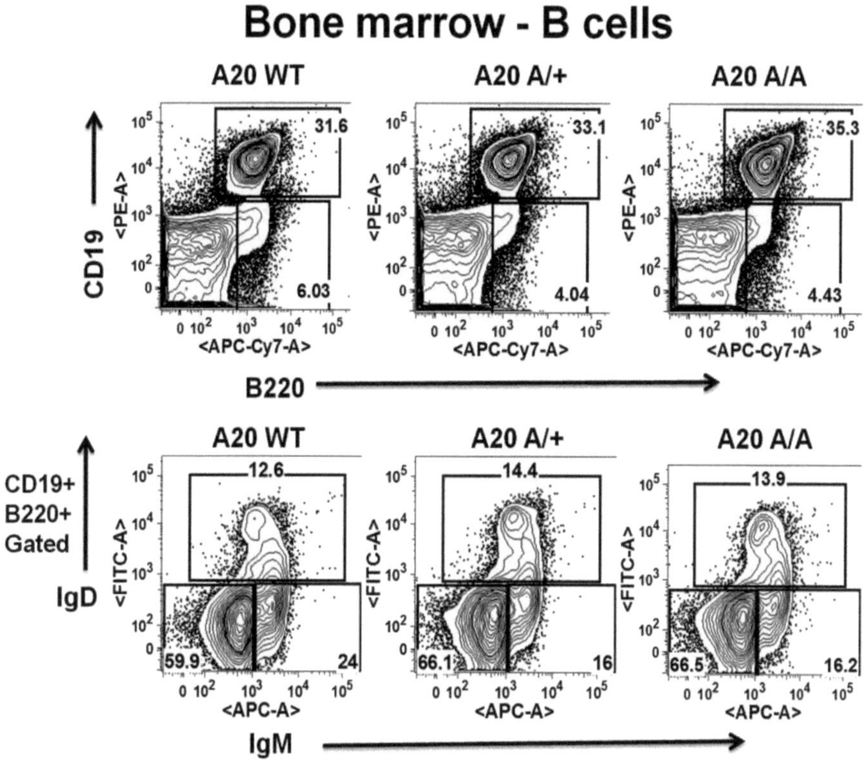

Fig. 3.2C Characterizing the B cells of the bone marrow (Copyright: Arnab De, Teruki Dainichi, Chozha Vendan Rathinam and Sankar Ghosh; previously published in EMBO Reports, 15(7), 2014, 775-783)

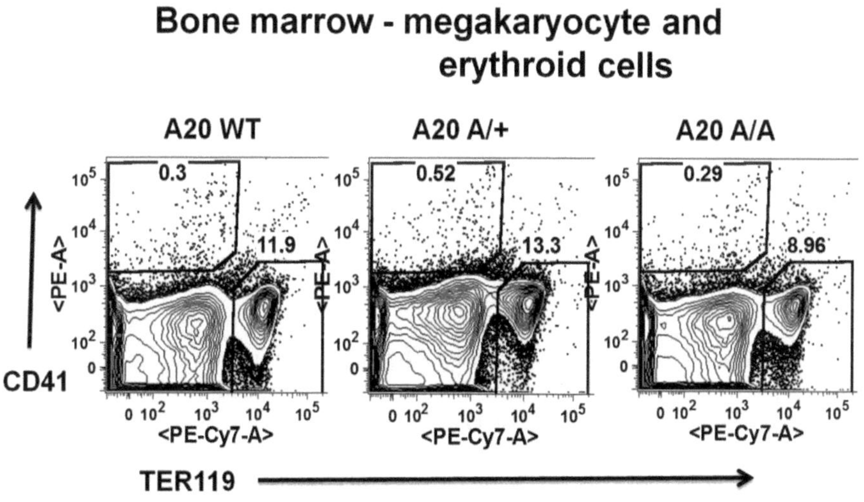

Fig. 3.2D Characterizing the megakaryocytes and erythroid cells of the bone marrow (Copyright: Arnab De, Teruki Dainichi, Chozha Vendan Rathinam and Sankar Ghosh; previously published in EMBO Reports, 15(7), 2014, 775-783)

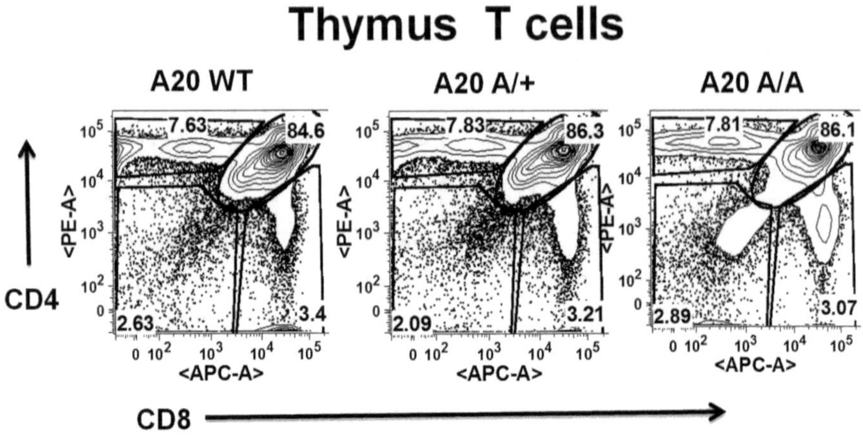

Fig. 3.2E Characterizing the CD4 and CD8 cell lineage of thymus (Copyright: Arnab De, Teruki Dainichi, Chozha Vendan Rathinam and Sankar Ghosh; previously published in EMBO Reports, 15(7), 2014, 775-783)

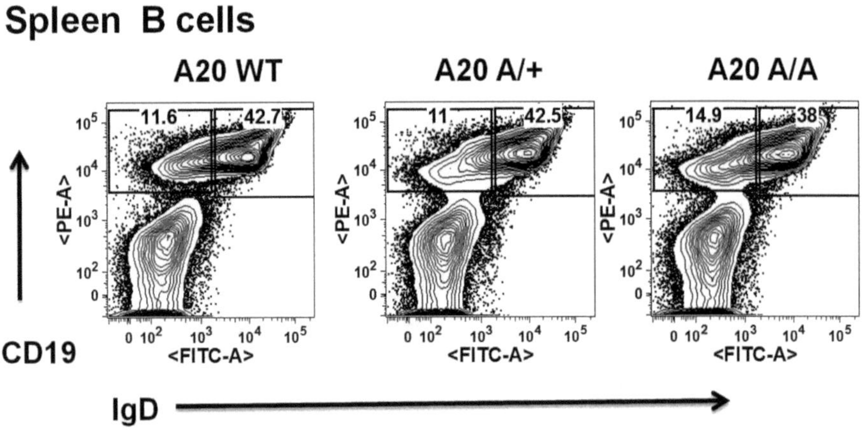

Fig. 3.2F Characterizing the B cells of the spleen (Copyright: Arnab De, Teruki Dainichi, Chozha Vendan Rathinam and Sankar Ghosh; previously published in EMBO Reports, 15(7), 2014, 775-783)

of either GMCSF or MCSF respectively. After 7 days of in vitro culture, we stimulated them with either TNFα or LPS. ELISA analysis revealed that LPS stimulation produced roughly equal amounts of TNFα, IL-6, and IL-12 by BMDM and BMDCs of both C/C and A/A mice (Fig. 3.3A). Similar results were obtained following TNFα stimulation (IL-6 and IL-12 were roughly similar for both C/C and A/A littermates) as shown in Fig. 3.3B.

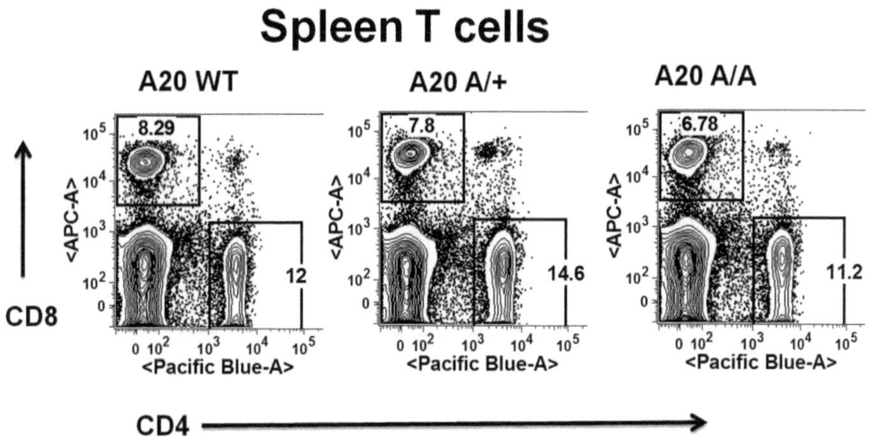

Fig. 3.2G Characterizing the T cells of the spleen (Copyright: Arnab De, Teruki Dainichi, Chozha Vendan Rathinam and Sankar Ghosh; previously published in EMBO Reports, 15(7), 2014, 775-783)

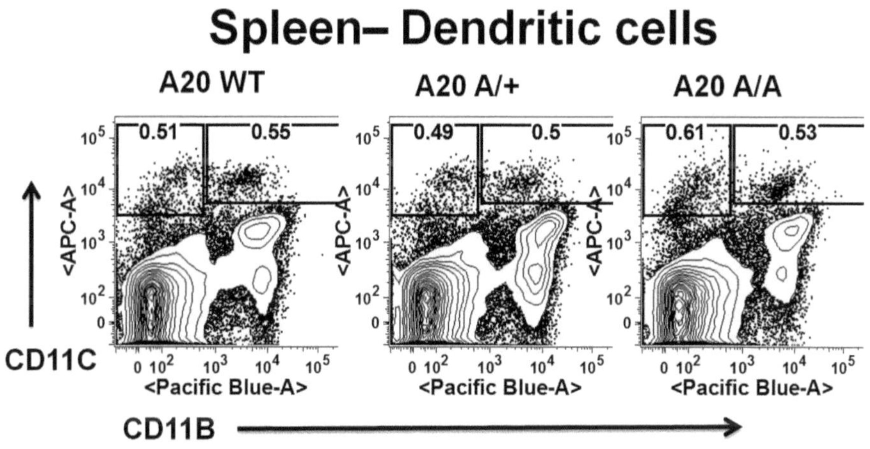

Fig. 3.2H Characterizing the dendritic cells of the spleen (Copyright: Arnab De, Teruki Dainichi, Chozha Vendan Rathinam and Sankar Ghosh; previously published in EMBO Reports, 15(7), 2014, 775-783)

In addition, we also analyzed the DCs and macrophages by flow cytometry for their activation status (as determined by upregulation of CD40, CD80, CD86, and MHC-Class II) after 48 h of stimulation with LPS and TNFα. While both DCs and macrophages showed upregulation (of CD40, CD80, CD86, and MHC-Class II) upon stimulation, the levels of expression were comparable amongst C/C, C/A, and A/A littermates (Figs. 3.3C and 3.3D).

Taken together, these data indicated that the C103A deubiquitinase domain is not involved in limiting inducible NF-κB activation.

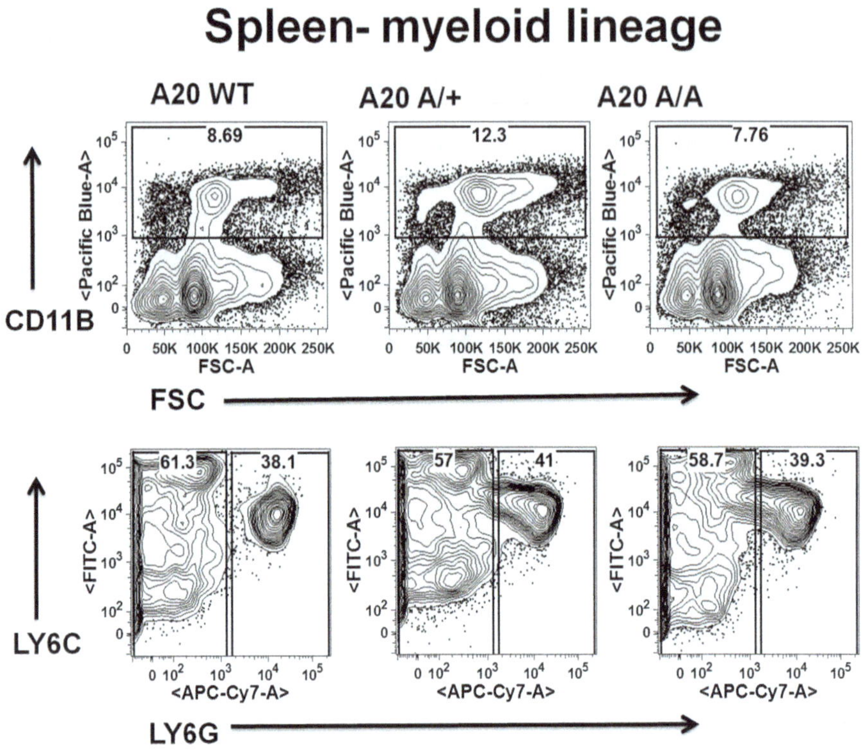

Fig. 3.2I Characterizing the myeloid cells of the spleen (Copyright: Arnab De, Teruki Dainichi, Chozha Vendan Rathinam and Sankar Ghosh; previously published in EMBO Reports, 15(7), 2014, 775-783)

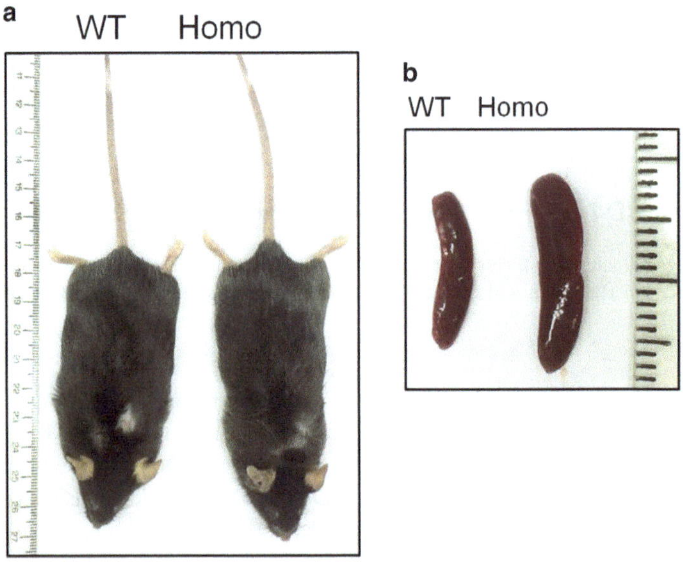

Fig. 3.2J (**a**) Photograph of 6-month-old C/C (WT) and A/A (Homo KI) mice (**b**) photograph of spleen of 6-month-old C/C (WT) and A/A (Homo KI) mice (Copyright: Arnab De, Teruki Dainichi, Chozha Vendan Rathinam and Sankar Ghosh; previously published in EMBO Reports, 15(7), 2014, 775-783)

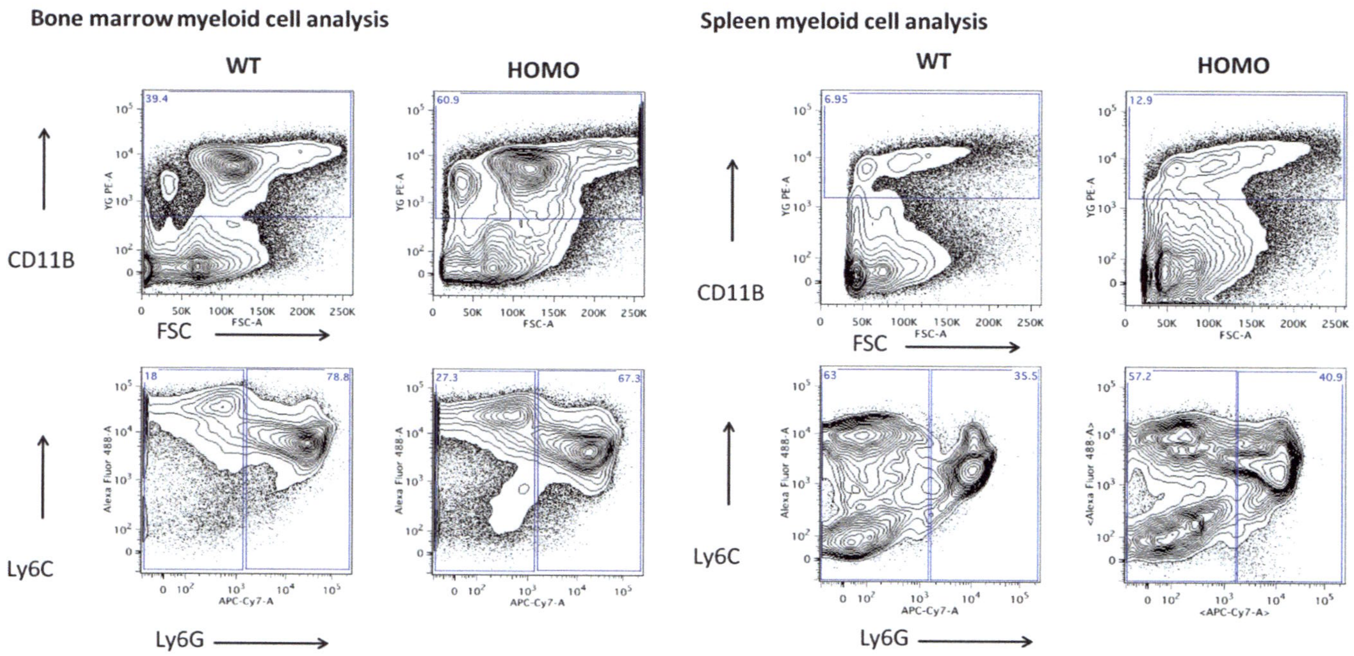

Fig. 3.2K Six-month-old A/A (Homo KI) mice show increased numbers of myeloid cells as compared to sex-matched C/C (wild-type) littermates (Copyright: Arnab De, Teruki Dainichi, Chozha Vendan Rathinam and Sankar Ghosh; previously published in EMBO Reports, 15(7), 2014, 775-783)

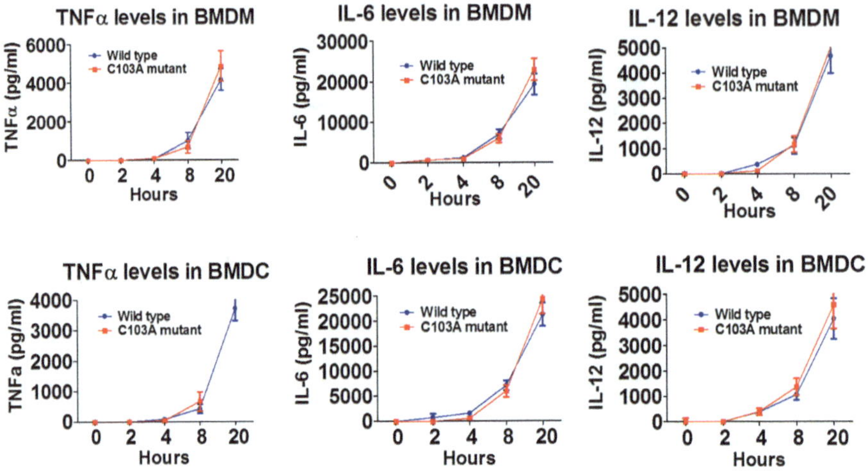

Fig. 3.3A TNFα, IL-6, and IL-12 produced by BMDM and BMDC in response to 1 μg/mL LPS was measured by ELISA. *Error bars* represent standard deviation (Copyright: Arnab De, Teruki Dainichi, Chozha Vendan Rathinam and Sankar Ghosh; previously published in EMBO Reports, 15(7), 2014, 775-783)

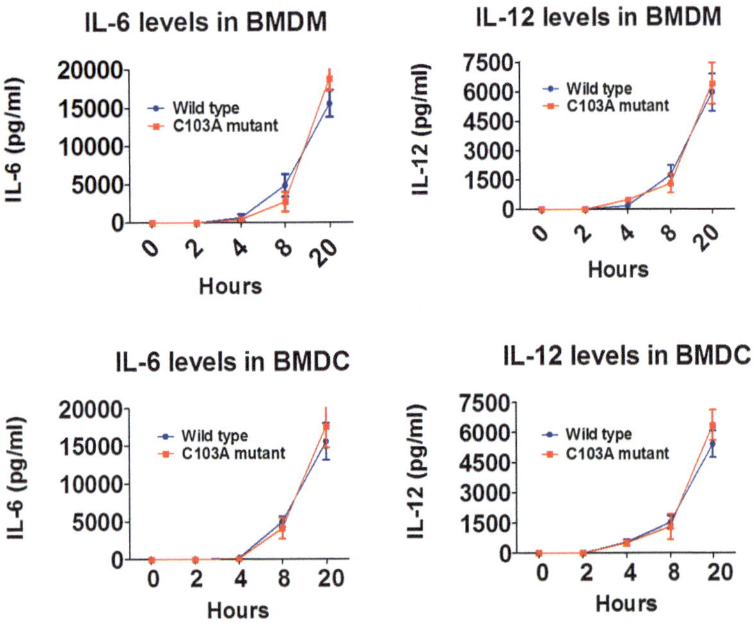

Fig. 3.3B IL-6 and IL-12 produced by BMDM and BMDC in response to 10 ng/mL TNFα was measured by ELISA. *Error bars* represent standard deviation (Copyright: Arnab De, Teruki Dainichi, Chozha Vendan Rathinam and Sankar Ghosh; previously published in EMBO Reports, 15(7), 2014, 775-783)

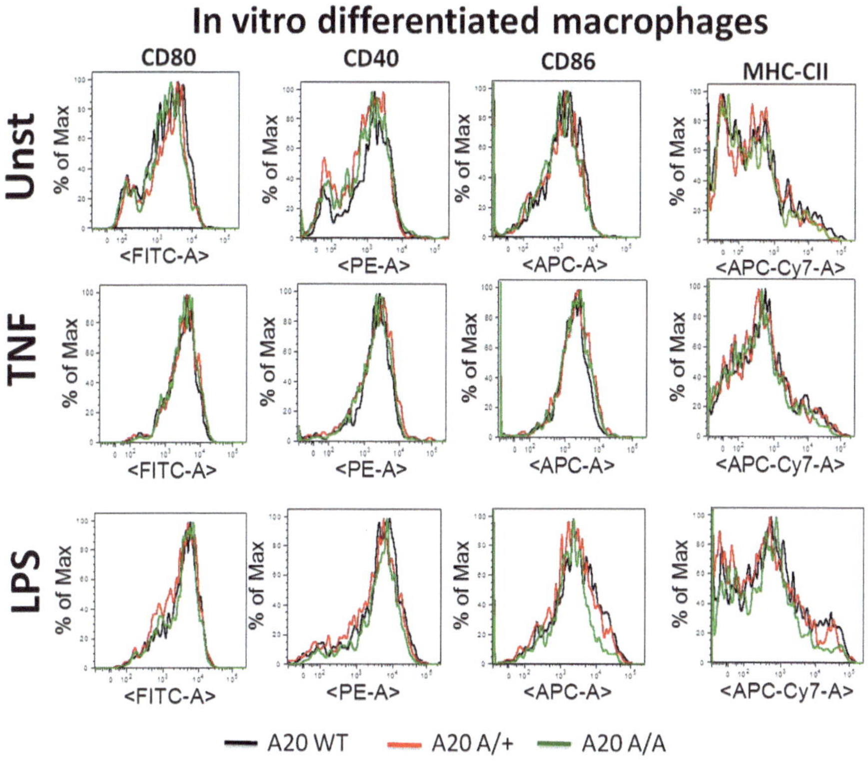

Fig. 3.3C Analysis of activation status of BMDM by flow cytometry (Copyright: Arnab De, Teruki Dainichi, Chozha Vendan Rathinam and Sankar Ghosh; previously published in EMBO Reports, 15(7), 2014, 775-783)

3.2.5 Response of A20 OTU Knock-In Mice to LPS Shock

The perinatal lethality manifested by lack of A20 is largely mediated by TLR-based signaling. Thus, A20 is important for restricting LPS-induced inflammation in vivo and we wanted to determine the physiological contribution of the deubiquitinase activity of A20 in LPS-dependent responses. We therefore challenged the C/C and A/A knock-in mice with LPS using established protocols and observed that both WT and the A/A knock-in mice succumbed to LPS shock (Fig. 3.3E). We also analyzed the serum levels of the key acute phase cytokines (TNFα, IL-6, and IL-12) after LPS injection and found that while TNFα peaked after 1 h, IL-6 and IL-12 peaked later in agreement with previous studies. However, there was no significant difference in the levels of the cytokines between WT and the A/A mice after LPS shock (Fig. 3.3F). Hence, the deubiquitinating function of A20 does not play an important role in limiting inflammatory responses in vivo following LPS administration.

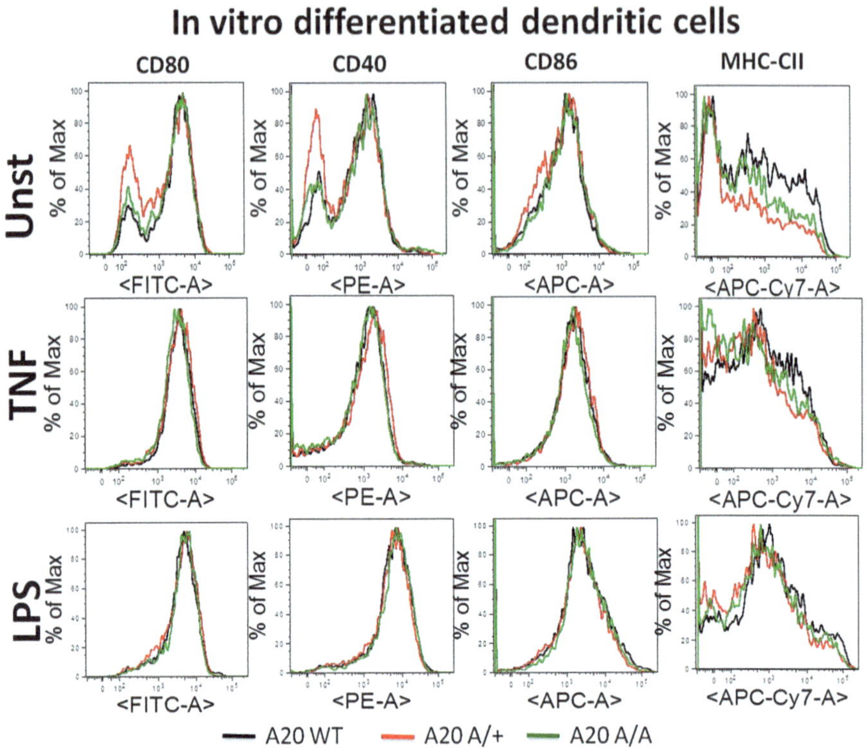

Fig. 3.3D Analysis of activation status of BMDC by flow cytometry (Copyright: Arnab De, Teruki Dainichi, Chozha Vendan Rathinam and Sankar Ghosh; previously published in EMBO Reports, 15(7), 2014, 775-783)

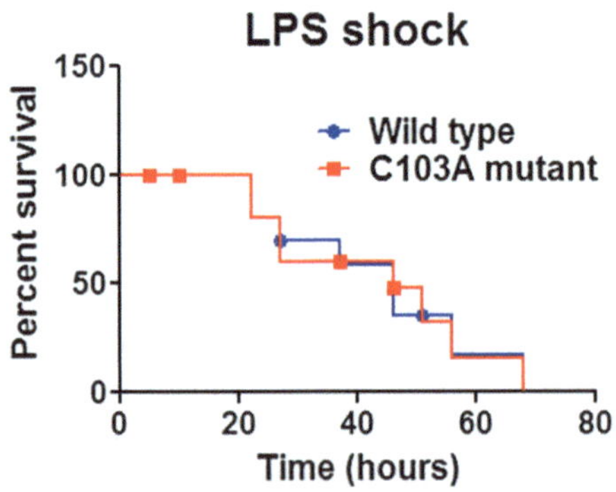

Fig. 3.3E Age- and sex-matched mice ($n=5$; experiment repeated thrice) were given intraperitonial injections of 50 μg/mL LPS and survival was scored every 6 h for 72 h (Copyright: Arnab De, Teruki Dainichi, Chozha Vendan Rathinam and Sankar Ghosh; previously published in EMBO Reports, 15(7), 2014, 775-783)

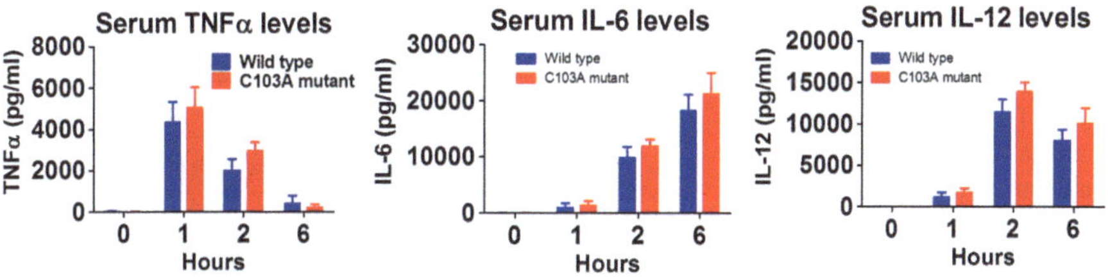

Fig. 3.3F Serum cytokine levels of TNFα, IL-6, and IL-12 in mice injected with 50 μg/mL LPS were measured by ELISA for indicated time points. *Error bars* represent standard deviation (Copyright: Arnab De, Teruki Dainichi, Chozha Vendan Rathinam and Sankar Ghosh; previously published in EMBO Reports, 15(7), 2014, 775-783)

3.2.6 Similar Activation of NF-κB in BMDMs from Both Wild-Type and Homozygous Mice

Before biochemically testing the activation of NF-κB in C/C and A/A littermates, we first did an in vitro deubiquitinase assay incubating recombinant K48- or K63-linked polyubiquitin chains with immunoprecipitated A20 analogs (wild-type and C103A-A20) from LPS-stimulated BMDMs of littermates and showed that the C103A mutation eliminates deubiquitinase activity in A20. Our results clearly demonstrate that the C103A mutation eliminates DUB activity of A20, as C103A-A20 is unable to deubiquitinate either K48-linked or K63-linked ubiquitin chains (Fig. 3.4A). However, wt-A20 can deubiquitinate both K48-linked and K63-linked ubiquitin chains in the in vitro DUB assay in agreement with previous results [17–19].

Next, we wanted to test the role of the A20 deubiquitinase activity in NF-κB activation in cells stimulated with TNFα. We isolated and cultured BMDMs from C/C and A/A knock-in mice, stimulated them with TNFα, and prepared nuclear fractions to test NF-κB binding to DNA by EMSA (Fig. 3.4B). Both the WT and the A/A BMDMs showed elevated (but similar) binding of p65-p50 heterodimers to the DNA following stimulation. The nuclear extract isolated after stimulating the WT BMDMs for 30 min was used for the super-shift assay and analysis with the unlabeled probe. Western blotting showed similar activation of NF-κB as tracked by degradation of IκBα. As RIP1 in stimulated cells has been suggested to be deubiquitinated by A20, we immunoprecipitated RIP1 from stimulated cells and tested it for ubiquitination. Both the WT and the C103A-A20 were recruited to RIP1 in roughly equal amounts and similar kinetics following TNFα stimulation (Fig. 3.4C). Intriguingly, we also found roughly equal amounts of ubiquitinated RIP1 in the BMDMs of both genotypes following TNFα stimulation. These results suggest that the deubiquitinase activity of A20 does not affect RIP1 ubiquitination in cells.

Since A20 has also been reported to inhibit TNF-induced cell death [3, 20], we treated BMDMs derived from C/C and A/A littermates with TNFα. Staining with Annexin V and propidium iodide showed that cell death is similar in both the

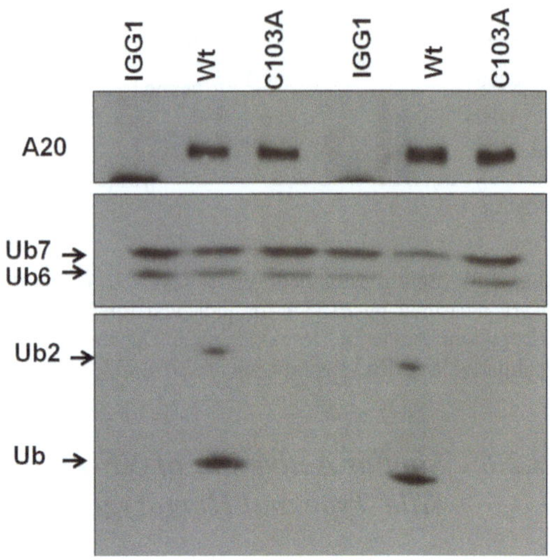

Fig. 3.4A C103A mutation eliminates deubiquitinase activity of A20 (Copyright: Arnab De, Teruki Dainichi, Chozha Vendan Rathinam and Sankar Ghosh; previously published in EMBO Reports, 15(7), 2014, 775-783)

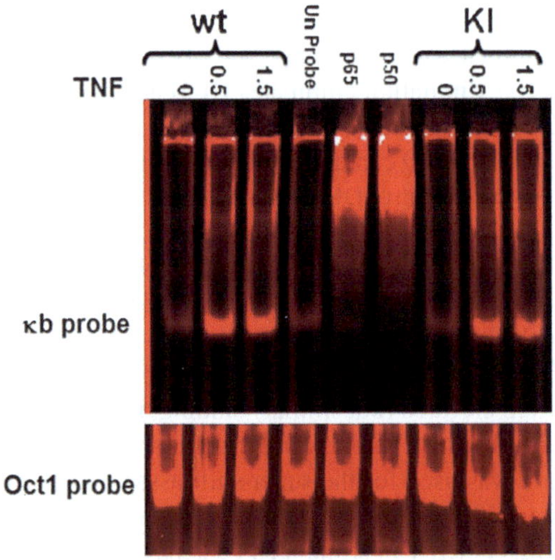

Fig. 3.4B NF-κB binding to DNA in response to TNFα stimulation was analyzed by Electrophoretic mobility shift assay. EMSA was performed with nuclear extracts after stimulating BMDMs isolated from wild-type and homozygous littermates with 10 ng/mL of TNFα for the indicated time points. The nuclear extract isolated at 30 min from wild-type BMDMs was used for the super-shift assay (with p65 and p50 antibodies) and analysis with the unlabelled κB probe (Copyright: Arnab De, Teruki Dainichi, Chozha Vendan Rathinam and Sankar Ghosh; previously published in EMBO Reports, 15(7), 2014, 775-783)

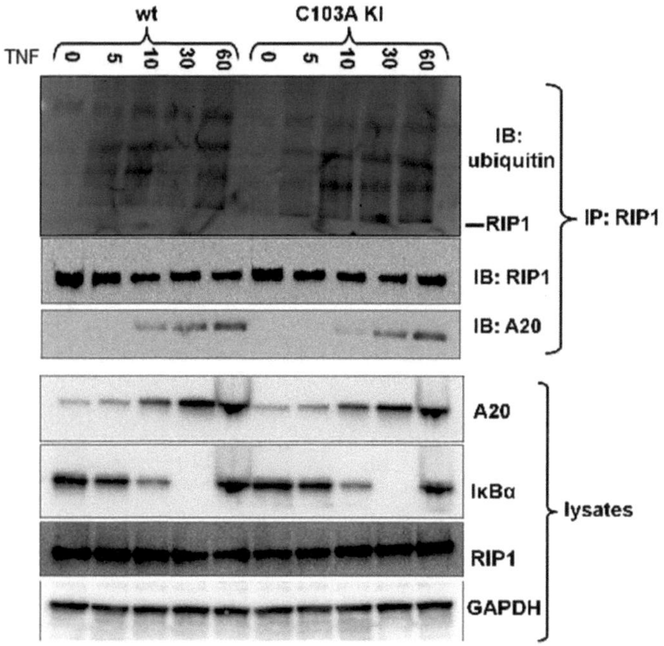

Fig. 3.4C BMDMs isolated from wild-type and homozygous littermates were stimulated with 10 ng/mL TNFα for the indicated time points. The cells were lysed and immunoprecipitated with RIP1 antibody and immunoblotted with indicated antibodies (Copyright: Arnab De, Teruki Dainichi, Chozha Vendan Rathinam and Sankar Ghosh; previously published in EMBO Reports, 15(7), 2014, 775-783)

genotypes (Fig. 3.4D). Hence, we conclude that the deubiquitinase activity of A20 is not responsible for inhibiting TNFα-induced cell death.

Similar results were also obtained upon stimulation of BMDMs with LPS. Both the wild-type and the homozygous mutant showed similar activation of NF-κB as analyzed by EMSA (Fig. 3.4E). The result is consistent with the kinetics of disappearance of IκBα as analyzed by western blotting (Fig. 3.4F).

3.3 Discussion

Ubiquitination is a reversible posttranslational modification that involves the attachment of one or more ubiquitin monomers to the substrate. This reversible posttranslational modification involves the activity of ubiquitinating enzymes (like kinases) that add ubiquitin chains to the substrate, and deubiquitinases (like phosphatases) that disassembles these ubiquitin chains. While ubiquitination is best known to target the substrate protein for proteasome-mediated degradation, recent

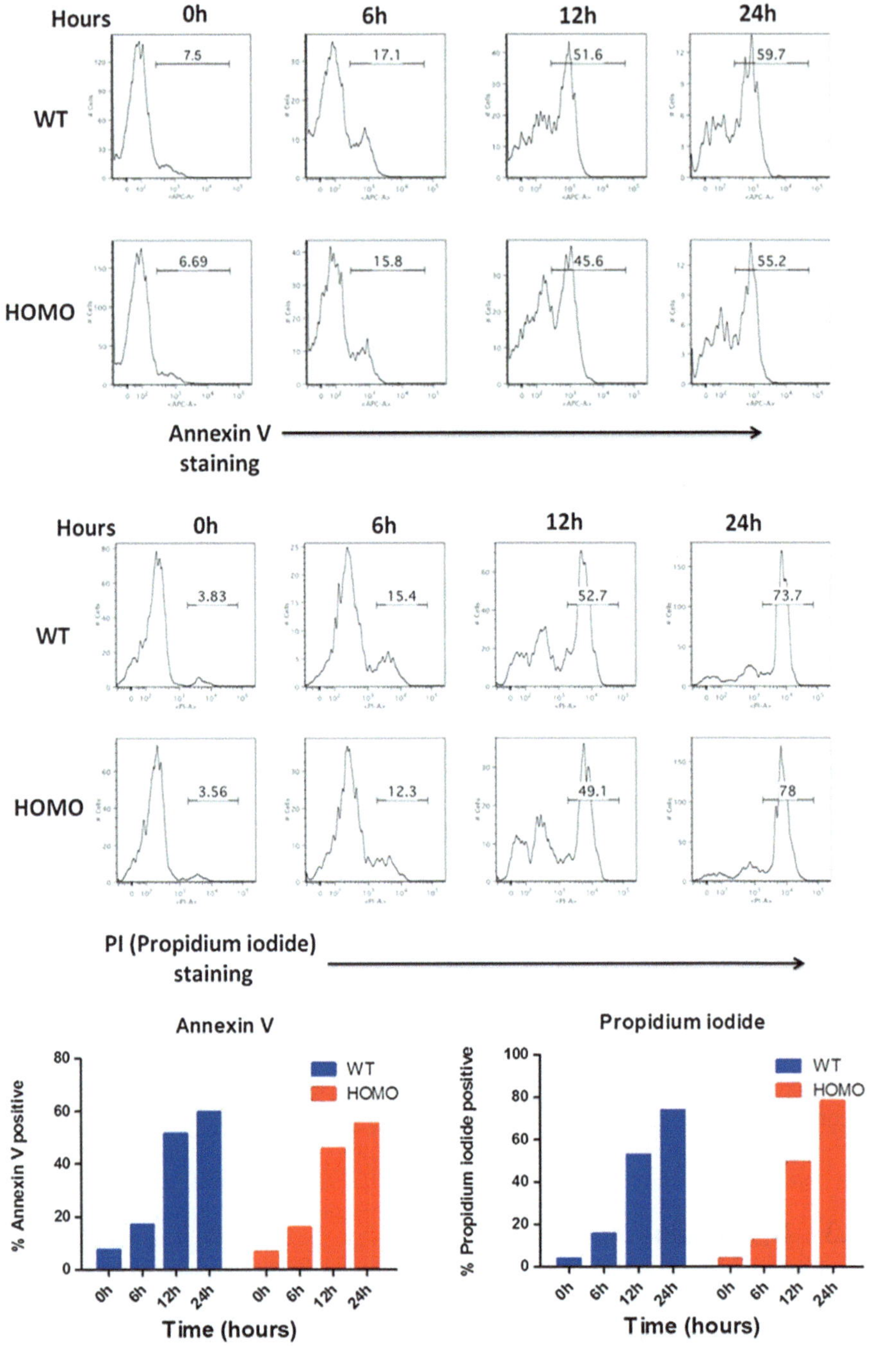

Fig. 3.4D Annexin V and propidium iodide staining of TNFα stimulated BMDMs derived from wild-type (C/C) and homozygous (A/A) littermates; bar graph showing percentage of cell death in TNFα stimulated BMDMs derived from wild-type and homozygous littermates (Copyright: Arnab De, Teruki Dainichi, Chozha Vendan Rathinam and Sankar Ghosh; previously published in EMBO Reports, 15(7), 2014, 775-783)

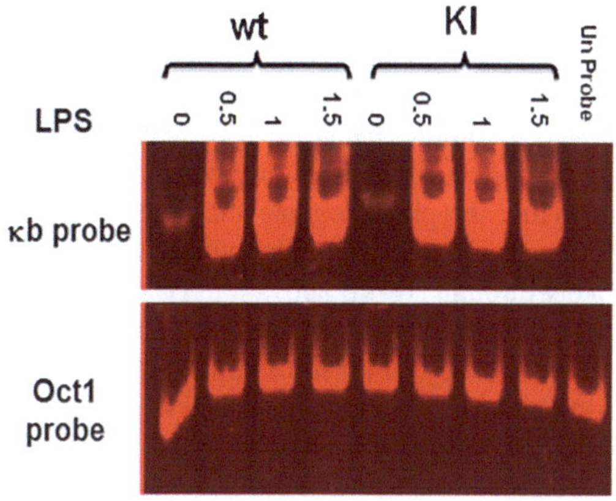

Fig. 3.4E EMSA was performed as in (A); stimulation was with 1 μg/mL of LPS (Copyright: Arnab De, Teruki Dainichi, Chozha Vendan Rathinam and Sankar Ghosh; previously published in EMBO Reports, 15(7), 2014, 775-783)

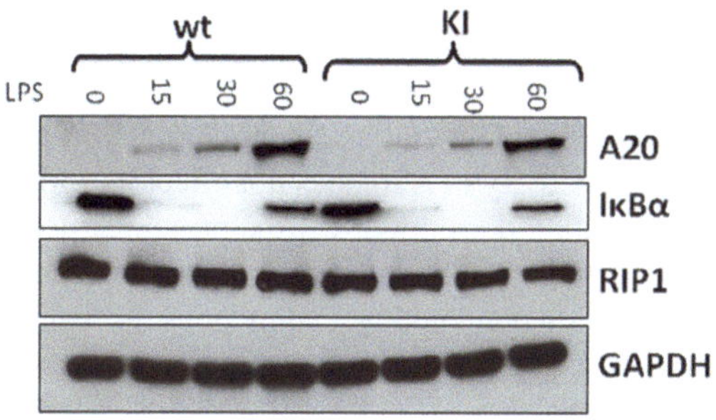

Fig. 3.4F BMDMs isolated from wild-type and homozygous littermates were stimulated with 1 μg/mL of LPS for the indicated time points. The cells were lysed and immunoblotted with indicated antibodies (Copyright: Arnab De, Teruki Dainichi, Chozha Vendan Rathinam and Sankar Ghosh; previously published in EMBO Reports, 15(7), 2014, 775-783)

studies have suggested various nonproteolytic consequences of ubiquitination. K48-linked ubiquitination of substrates targets them for degradation, while regulatory K63-linked ubiquitination has been shown to activate various kinases including IKK in vitro leading to activation of NF-κB. Thus ubiquitination has emerged as a mechanism whereby covalent attachment of diverse polyubiquitin chains lead to different biological outcomes. However while it is clear that K48-linked polyubiquitination leads to proteasomal degradation, the physiological consequences of regulatory K63-linked ubiquitination remains to be established in animal models [21].

A20 is one of the best characterized of all enzymes that acts on K63-linked ubiquitin chains. It is believed to deubiquitinate K63-linked ubiquitin chains and this deubiquitinase activity has been proposed to be essential in the downregulation of NF-κB signaling for both TNFα- and LPS-mediated pathways. However, this is difficult to reconcile with the observation that the OTU domain of A20 disassembles

K48-linked polyubiquitin chains much more potently as compared to K63-linked polyubiquitin chains in an in vitro setting [17, 18]. This is an intriguing discrepancy as the deubiquitinating function of A20 would be expected to be specific to K63-linked ubiquitin chains for it to play such a profound role in a tightly regulated process such as NF-κB activation.

While the physiological role of the deubiquitinating property of A20 requires further study, there is no doubt that A20 is essential in limiting inflammation, very likely through NF-κB activation. Dysregulation of A20 has now been implicated in various autoimmune diseases and cancer. Polymorphisms in the A20 locus increase disease susceptibility in multiple autoimmune diseases including type I diabetes, psoriasis, rheumatoid, arthritis, systemic lupus erythematosus, etc. [15]. Recently, A20 polymorphisms (A125V and F127C) were discovered in the DUB domain, which increases susceptibility to autoimmunity by impairing A20-mediated deubiquitination [22]. It has been speculated that the Cys103, the residue primary important for deubiquitinating property could be important in the physiological mechanism of A20.

To directly ascertain the contribution of the deubiquitinating role of A20 in its biological function, we replaced the wild-type A20 gene with a mutant form that lacks deubiquitinating activity. Our studies of these knock-in mice show that abolishing the deubiquitinase activity of A20 does not affect its ability to be recruited to RIP1, and hence the mutation does not affect the protein structure. Furthermore, it is important to appreciate that the phenotypes observed in the knock-in mice are the result of the C103A mutation regulated by endogenous mechanisms as compared to previous in vitro experiments that relied on overexpression systems and exogenous gene regulatory mechanisms.

The mutant protein in the homozygous mouse and the wild-type protein in the wild-type littermate are induced in equal amounts following inflammatory stimuli. This is what is typically expected in a gene-targeted knock-in system. However, as A20 expression is directly induced by NF-κB signaling, A20's capacity to downregulate NF-κB signals exemplifies a negative feedback mechanism. Moreover, previous studies of hypomorphic $A20^{+/-}$ cells suggest that the ability of A20 to negatively regulate NF-κB signaling is directly dependent on the amount of A20 in the cells [23, 24]. Thus if Cys103 was indeed important for restricting NF-κB signaling, it would be expected that the levels of the mutant protein would be increased in the homozygous mice. This was the first hint that Cys103 may not be playing an important role in downregulating NF-κB signaling in a physiological context. This is not surprising as the deubiquitinase activity mediated by Cys103 has been shown to promiscuously cleave unanchored K11-, K48-, and K63-linked polyubiquitin chains in addition to disassembling K63-linked chains from in vitro substrates [8, 11, 17–19].

A20's Cys103 residue has been proposed to be crucial for downregulating NF-κB signaling in vitro [11, 13]. Thus, it was expected that the A20-C103A mice would at least partially resemble the $A20^{-/-}$ mice [7]. However, while $A20^{-/-}$ mice develop spontaneous multiorgan inflammation and perinatal lethality, the A20-C103A mice are grossly normal in the basal state. While $A20^{-/-}$ mice have an increased number of myeloid lineage cells in the bone marrow and spleen, we found an overall normal population of both myeloid and lymphoid cells in the bone marrow, spleen, and thymus in the C/C, C/A, and A/A littermates. Thus, the deubiquitinating activity of A20 does not play a major role in the maintenance of basal immune homeostasis. Moreover even following LPS stimulation, the response of the AA knock-ins was similar to wild-type littermates.

Our results are in agreement with a recent study that also found that mice lacking the deubiquitinating activity of A20 are grossly normal for at least 4 months and contained normal number of lymphocytes [10]. These investigators also concluded, similar to our observations, that the deubiquitinase activity of A20 was not required for prevention of spontaneous cachexia and premature death. Surprisingly however, this study focused exclusively on the role of Cys103 in limiting TNFα-induced NF-κB activation, despite previous reports from the same group that A20 functioned in vivo in limiting TLR/MyD88-dependent pathways [8, 9]. Hence, our study focusing on the role of A20 C103 on LPS/TLR-mediated responses is vital for understanding the physiological role of the deubiquitinating property of A20 in restricting persistent TLR-mediated activation of NF-κB.

As our studies clearly demonstrate, the deubiquitinase activity of A20 is dispensable for its regulation of NF-κB signaling. However, the E3 ligase function of the ZnF4 domain could partially compensate for the lack of deubiquitinating function in vivo, even though A20 is only supposed to ubiquitinate substrates such as RIP1 after first deubiquitinating them. Interesting, a recent study with mice lacking the E3 ligase activity demonstrated that neither of the C103 or the ZnF4 motif were singly responsible for all of A20's functions in restricting TNFα signaling, and in the ZnF4 motif appeared more important than the deubiquitinase activity in restricting TNFα signaling in embryonic fibroblasts (MEFs). The mechanism underlying such an observation remains unclear, but also casts doubt on the importance of A20 as a deubiquitinating enzyme. As we know that the A20 protein is important in inflammatory signaling, these studies suggest that A20 plays a more important role that does not involve ubiquitination/deubiquitination in exerting its regulatory function in inflammatory signaling. In this respect, the seventh zinc-finger motif (ZnF7) of A20 was recently proposed to be involved in direct inhibition of IKK by a noncatalytic mechanism [25]. Thus, it is probably fair to say that significant additional studies will be needed to determine the actual role that A20 plays in regulating inflammatory/NF-κB signaling.

3.4 Materials and Methods

3.4.1 Generation of A20C103A Knock-In Mice

The A20C103A mice were generated using conventional gene-targeting approaches. In brief, exon 3 harboring Cys103 was PCRed out using a bacterial artificial chromosome (BAC) bearing the A20 gene and cloned into a PL452 plasmid having a

Lox-Neo-Lox cassette. Standard site-directed mutagenesis using Stratagene kits was used to mutate Cys103 to Alanine. The relevant part of the A20 construct was then transferred from PL452 plasmid to the BAC using homologous recombination. This BAC construct which now harbors the Lox-Ne0-Lox cassette along with a C103A mutation was retrieved into a pMCS_DTA plasmid (with diphtheria toxin selection gene) which was linearized and electroporated into CSL2J2, albino C57BL/6J ES cells. Correctly screened ES cell clones were injected into blastocysts derived from C57BL/6 mice to give rise to chimeras (in the Columbia University Transgenic Core facility). Genetic transmission of the allele was confirmed by PCR and subsequent sequencing to confirm the presence of the C103A mutation. The LoxP flanked neomycin sequences were deleted by crossing the transgenic mice with the EIIA-Cre deleter mice. The C/A heterozygous mice were bred to generate age-matched C/C wild-type and A/A homozygous mice for various experiments.

3.4.2 Cells

Bone marrow cells of 6–8 weeks age-matched wild-type (C/C), heterozygous (C/A) and homozygous (A/A) mice were differentiated into dendritic cells and macrophages in the presence of either GMCSF or MCSF respectively. In the in vitro experiments, BMDMs were stimulated with 10 ng/mL TNFα or 1 μg/mL LPS.

3.4.3 Biochemical Experiments

Immunoprecipitations and western blotting were performed as has been described previously [11]. Gel shift assay was done using the Li-cor EMSA kit as per the user manual. The κB probe was purchased from Li-cor while the Oct1 probe was custom-made. Antibodies used in this study include anti-murine RIP monoclonals (BD Bioscience, clone 610458), ubiquitin (Santa Cruz, sc-8017), IκBα (Santa Cruz, sc-371), A20 (Imgenex, 161A), βtubulin (Abcam).

3.4.4 LPS-Induced Shock

LPS was injected intraperitonially at a concentration of 50 mg kg^{-1} of mice body weight. The mice were monitored for survival every 8 h. In a separate experiment, the mice were bled 1, 2, and 6 h after LPS treatment and the serum cytokine levels measured by ELISA [26].

3.4.5 *Flow Cytometry, ELISA and qRT–PCR*

Cell preparations, flow cytometric and ELISA analyses were performed as previously described [24, 26]. ELISA was performed with kits from BD Biosciences. Cells were analyzed by flow cytometry using LSRII and Flowjo software (Tree Star).

BMDMs were stimulated and RNA was isolated using the RNA Easy kit (Qiagen). RNA was reverse transcribed (SuperScriptIII reverse transcriptase; Invitrogen-Life Technologies), and SYBR Green master mix (QuantiTect SYBR green; Invotrogen) was used to quantify relative gene expression of the corresponding mRNA with normalization to β-actin [26].

References

1. Hayden MS, Ghosh S (2012) NF-kappaB, the first quarter-century: remarkable progress and outstanding questions. Genes Dev 26:203–234
2. Chen ZJ, Sun LJ (2009) Nonproteolytic functions of ubiquitin in cell signaling. Mol Cell 33:275–286
3. Opipari AW Jr, Hu HM, Yabkowitz R, Dixit VM (1992) The A20 zinc finger protein protects cells from tumor necrosis factor cytotoxicity. J Biol Chem 267:12424–12427
4. Song HY, Rothe M, Goeddel DV (1996) The tumor necrosis factor-inducible zinc finger protein A20 interacts with TRAF1/TRAF2 and inhibits NF-kappaB activation. Proc Natl Acad Sci U S A 93:6721–6725
5. Heyninck K, De Valck D, Vanden Berghe W, Van Criekinge W, Contreras R, Fiers W, Haegeman G, Beyaert R (1999) The zinc finger protein A20 inhibits TNF-induced NF-kappaB-dependent gene expression by interfering with an RIP- or TRAF2-mediated transactivation signal and directly binds to a novel NF-kappaB-inhibiting protein ABIN. J Cell Biol 145:1471–1482
6. Jaattela M, Mouritzen H, Elling F, Bastholm L (1996) A20 zinc finger protein inhibits TNF and IL-1 signaling. J Immunol 156:1166–1173
7. Lee EG, Boone DL, Chai S, Libby SL, Chien M, Lodolce JP, Ma A (2000) Failure to regulate TNF-induced NF-κB and cell death responses in A20-deficient mice. Science 289:2350–2354
8. Boone DL et al (2004) The ubiquitin-modifying enzyme A20 is required for termination of Toll-like receptor responses. Nat Immunol 5:1052–1060
9. Turer EE, Tavares RM, Mortier E, Hitotsumatsu O, Advincula R, Lee B, Shifrin N, Malynn BA, Ma A (2008) Homeostatic MyD88-dependent signals cause lethal inflammation in the absence of A20. J Exp Med 205:451–464
10. Lu TT et al (2013) Dimerization and ubiquitin mediated recruitment of A20, a complex deubiquitinating enzyme. Immunity 38:896–905
11. Wertz IE et al (2004) De-ubiquitination and ubiquitin ligase domains of A20 downregulate NF-kappaB signalling. Nature 430:694–699
12. Evans PC, Ovaa H, Hamon M, Kilshaw PJ, Hamm S, Bauer S, Ploegh HL, Smith TS (2004) Zinc-finger protein A20, a regulator of inflammation and cell survival, has de-ubiquitinating activity. Biochem J 378:727–734
13. Shembade N, Ma A, Harhaj EW (2010) Inhibition of NF-kappaB signaling by A20 through disruption of ubiquitin enzyme complexes. Science 327:1135–1139

14. Li L, Hailey DW, Soetandyo N, Li W, Lippincott-Schwartz J, Shu HB, Ye Y (2008) Localization of A20 to a lysosome-associated compartment and its role in NFkappaB signaling. Biochim Biophys Acta 1783:1140–1149
15. Ma A, Malynn BA (2012) A20: linking a complex regulator of ubiquitylation to immunity and human disease. Nat Rev Immunol 12:774–785
16. Singer A, Adoro S, Park J-H (2008) Lineage fate and intense debate: myths, models and mechanisms of CD4- versus CD8-lineage choice. Nat Rev Immunol 8:788–801
17. Komander D, Barford D (2008) Structure of the A20 OTU domain and mechanistic insights into deubiquitination. Biochem J 409:77–85
18. Lin SC, Chung JY, Lamothe B, Rajashankar K, Lu M, Lo YC, Lam AY, Darnay BG, Wu H (2008) Molecular basis for the unique deubiquitinating activity of the NF-kappaB inhibitor A20. J Mol Biol 376:526–540
19. Bosanac I et al (2010) Ubiquitin binding to A20 ZnF4 is required for modulation of NF-kappaB signaling. Mol Cell 40:548–557
20. He KL, Ting AT (2002) A20 inhibits tumor necrosis factor (TNF) alpha-induced apoptosis by disrupting recruitment of TRADD and RIP to the TNF receptor 1 complex in Jurkat T cells. Mol Cell Biol 22:6034–6045
21. Liu S, Chen ZJ (2011) Expanding role of ubiquitination in NF-[kappa]B signaling. Cell Res 21:6–21
22. Lodolce JP et al (2010) African-derived genetic polymorphisms in TNFAIP3 mediate risk for autoimmunity. J Immunol 184:7001–7009
23. Werner SL, Kearns JD, Zadorozhnaya V, Lynch C, O'Dea E, Boldin MP, Ma A, Baltimore D, Hoffmann A (2008) Encoding NF-kappaB temporal control in response to TNF: distinct roles for the negative regulators IkappaBalpha and A20. Genes Dev 22:2093–2101
24. Tavares RM et al (2010) The ubiquitin modifying enzyme A20 restricts B cell survival and prevents autoimmunity. Immunity 33:181–191
25. Skaug B, Chen J, Du F, He J, Ma A, Chen ZJ (2011) Direct, noncatalytic mechanism of IKK inhibition by A20. Mol Cell 44:559–571
26. Rao P et al (2010) IkappaBbeta acts to inhibit and activate gene expression during the inflammatory response. Nature 466:1115–1119

Chapter 4
Perspectives on Regulatory Ubiquitination

It is well-appreciated that almost all NF-κB signaling pathways proceed via the activation of the IKK complex as NF-κB activity is absent in mice deficient in both IKKα and IKKβ [1]. Phosphorylation of Ser-177 and Ser-181 in the "active loop" of IKKβ is required for the activation of IKK [2]. However, how this phosphorylation happens is unknown [3]. There are two obvious possibilities: IKK could either autophosphorylate itself or be phosphorylated by some other kinase (like TAK1). The biggest issue is that the receptors leading to IKK activation lack kinase activity or other known enzymatic activity. Hence, either kinase activity needs to be recruited to the pathway or conformational changes upon receptor ligation must somehow lead to IKK phosphorylation.

It is conceivable that these conformational changes might be caused by the oligomerization of adaptor proteins upon receptor ligation. Another hypothesis is that the adaptor proteins may be ubiquitinated with K63-linked, non-degradative, ubiquitin chains; and it is these ubiquitin chains that activate TAK1/IKK. The thesis broadly tests the second hypothesis to determine if regulatory ubiquitination might lead to NF-κB activation.

Initially, it was thought that these ubiquitin chains provide the oligomeric structure for stabilizing the signaling complexes or recruits the kinase complexes, like the IKK complex (via NBD of NEMO) or TAK1 complex (via the NBD of the TAB adaptor proteins), bringing these kinases closer together for *trans*-phosphorylation. Another nonexclusive model holds that ubiquitination of upstream adapters may directly activate the downstream kinase complexes. Recently though, it was shown that even ubiquitin chains that are not anchored to any substrate can directly activate IKK or TAK1 [4]. If direct activation of kinase complexes by substrate-linked ubiquitin or free-ubiquitin were to occur, it would be expected that short ubiquitin chains would be as effective as long ubiquitin chain in activating the kinase complex. However, this was not found to be the case. Moreover, it is unclear as to how the IKK complex or the TAK/TAB complex would differentiate between regulatory

A. De, *Ubiquitin Chains: Degradation and Beyond*, Springer Theses,
DOI 10.1007/978-3-319-14965-3_4

polyubiquitin chains linked to one substrate from those linked to a different substrate (or differentiate the substrate-linked chains from unanchored ubiquitin chains).

Crucially, if ubiquitin chains were to directly activate the IKK complex, it would mean that IKKβ or NEMO would have to be ubiquitinated. In tissue-culture-based experiments and knockdown models, regulatory ubiquitination of NEMO at K-399 has been shown to be essential for NF-κB activation [5]. However, subsequently knock-in mice generated with a single mutation at this residue was found to have normal NF-κB activation. Hence, the genetic murine model has conclusively demonstrated that ubiquitination of NEMO at K-399 is not essential for NF-κB activation.

There is however a plethora of biochemical evidence to suggest that upstream adaptor molecules like RIP1 indeed get ubiquitinated in response to TNFα stimulation. Western-blot analysis shows the slower-migrating forms of RIP1, or ubiquitinated-RIP1 [6]. The slower-migrating forms of RIP1 are augmented if only RIP1 that is recruited to TNF receptor complex is analyzed [7, 8]. The ubiquitin linkage on RIP1 may be K63-linked [7, 9–11] or linear [8]. Moreover, RIP1 is essential for activation of NF-κB as concluded from mice deficient in RIP1 [12]. In spite of this, for reasons outlined below (in addition to my findings), it is doubtful if the ubiquitination of RIP1 is actually essential for RIP1-mediated activation of NF-κB.

RIP1 has been shown to be ubiquitinated on Lys-377 and reconstitution studies using the K377R mutant in cells deficient in RIP1 showed the importance of this residue in ubiquitination of RIP1. However, the K377R mutant is also not recruited to the activated TNF receptor [10, 11]. Hence, the implications of ubiquitination at this residue remain unclear. Ubc5 (E2) is important for RIP1 ubiquitination, and Ubc5 knockdown virtually abrogates RIP1 ubiquitination [13] in cells. The cells lacking UbC5 though have only modest impairment in IKK activation. Similarly TRAF2/5 (E3-ligase) is important for RIP1 ubiquitination. But reconstitution of cells deficient in both TRAF2 and TRAF5 with TRAF2 RING mutants abolishes RIP1 ubiquitination without affecting IKK activation [14]. Perhaps the cleanest in vitro experiment, replacing endogenous ubiquitin with K63R-ubiquitin (genome knock-in at the ubiquitin locus), has showed that K63-linked ubiquitin is not required for TNFR-mediated IKK activation [13].

The greatest support for regulatory ubiquitination actually comes from the well-established roles of deubiquitinases as negative regulators of IKK activation [15]. A20 plays an essential role in downregulation NF-κB activation and consequent inflammation, as mice lacking A20 die from persistent inflammation in multiple organs and cachexia [16]. In humans, dysregulation of A20 has been associated with multiple autoimmune disorders. However, what was not known till now was if the deubiquitinase function of A20 (directed mainly towards RIP1) was actually essential for downregulation of NF-κB.

The thesis shows that "the deubiquitinase activity of A20 is dispensable for its role in NF-κB signaling." The knock-in mouse that we have generated selectively abrogates the deubiquitinase function of A20 in vivo. These mice display normal NF-κB activation without showing any inflammatory phenotype. Our results clearly demonstrate that the well-characterized role of A20 in limiting inflammatory responses is due to effects other than deubiquitination of K63-ubiquitin chains. Given that the presumed role of A20 as a deubiquitinase has been used to support the importance of regulatory K63 ubiquitination in NF-κB signaling, we believe that our studies will help focus future research efforts into alternative target pathways that do not depend on K63 ubiquitination. As A20 clearly has an important role in many diseases, understanding the true molecular function of A20 remains an important and worthwhile goal for the future.

It is tempting to speculate that K63-linked ubiquitin chains may play an alternative role in stabilizing upstream signal complexes by preventing untimely and inadvertent K48-linked degradative ubiquitination of these signaling complexes. In this regard, both A20 and CYLD have been shown to work by not only disassembling K63-linked ubiquitin chains (deubiquitinase role) but also facilitating the assembly of K48-linked ubiquitin chains to their substrates [17, 18]. It might well be that the timely ligation of the well-characterized, K48-linked degradative ubiquitin chains is the function of regulatory ubiquitination.

In the bigger context, the crucial issue is determining the mechanism for activation of kinases downstream of receptors that lack inherent kinase activity. Given the absence of receptor-enzymatic activity, ubiquitination of adaptor proteins provides an appealing mechanism for the activation of IKK. However, extensive genetic experiments using knock-in models are required to show the importance of regulatory ubiquitination in the activation of NF-κB.

References

1. Li Q, Estepa G, Memet S, Israel A, Verma IM (2000) Complete lack of NF-κB activity in IKK1 and IKK2 double-deficient mice: additional defect in neurulation. Genes Dev 14:1729–1733
2. Delhase M, Hayakawa M, Chen Y, Karin M (1999) Positive and negative regulation of IkappaB kinase activity through IKKbeta subunit phosphorylation. Science 284:309–313
3. Hayden MS, Ghosh S (2012) NF-kappaB, the first quarter-century: remarkable progress and outstanding questions. Genes Dev 26:203–234
4. Xia ZP, Sun L, Chen X, Pineda G, Jiang X, Adhikari A, Zeng W, Chen ZJ (2009) Direct activation of protein kinases by unanchored polyubiquitin chains. Nature 461:114–119
5. Zhou H, Wertz I, O'Rourke K, Ultsch M, Seshagiri S, Eby M, Xiao W, Dixit VM (2004) Bcl10 activates the NF-kappaB pathway through ubiquitination of NEMO. Nature 427:167–171
6. Lee TH, Shank J, Cusson N, Kelliher MA (2004) The kinase activity of Rip1 is not required for tumor necrosis factor-alpha-induced IkappaB kinase or p38 MAP kinase activation or for the ubiquitination of Rip1 by Traf2. J Biol Chem 279:33185–33191
7. Wu CJ, Conze DB, Li T, Srinivasula SM, Ashwell JD (2006) Sensing of Lys 63-linked polyubiquitination by NEMO is a key event in NF-kappaB activation [corrected]. Nat Cell Biol 8:398–406
8. Haas TL et al (2009) Recruitment of the linear ubiquitin chain assembly complex stabilizes the TNF-R1 signaling complex and is required for TNF-mediated gene induction. Mol Cell 36:831–844
9. Cusson-Hermance N, Khurana S, Lee TH, Fitzgerald KA, Kelliher MA (2005) Rip1 mediates the Trif-dependent Toll-like receptor 3- and 4-induced NF-κB activation but does not contribute to interferon regulatory factor 3 activation. J Biol Chem 280:36560–36566

10. Ea CK, Deng L, Xia ZP, Pineda G, Chen ZJ (2006) Activation of IKK by TNFalpha requires site-specific ubiquitination of RIP1 and polyubiquitin binding by NEMO. Mol Cell 22:245–257
11. Li H, Kobayashi M, Blonska M, You Y, Lin X (2006) Ubiquitination of RIP is required for tumor necrosis factor α-induced NF-κB activation. J Biol Chem 281:13636–13643
12. Kelliher MA, Grimm S, Ishida Y, Kuo F, Stanger BZ, Leder P (1998) The death domain kinase RIP mediates the TNF-induced NF-kappaB signal. Immunity 8:297–303
13. Xu M, Skaug B, Zeng W, Chen ZJ (2009) A ubiquitin replacement strategy in human cells reveals distinct mechanisms of IKK activation by TNFalpha and IL-1beta. Mol Cell 36:302–314
14. Zhang L, Blackwell K, Shi Z, Habelhah H (2010) The RING domain of TRAF2 plays an essential role in the inhibition of TNFalpha-induced cell death but not in the activation of NF-kappaB. J Mol Biol 396:528–539
15. Harhaj EW, Dixit VM (2011) Deubiquitinases in the regulation of NF-[kappa]B signaling. Cell Res 21:22–39
16. Lee EG, Boone DL, Chai S, Libby SL, Chien M, Lodolce JP, Ma A (2000) Failure to regulate TNF-induced NF-κB and cell death responses in A20-deficient mice. Science 289:2350–2354
17. Wertz IE et al (2004) De-ubiquitination and ubiquitin ligase domains of A20 downregulate NF-kappaB signalling. Nature 430:694–699
18. Ahmed N, Zeng M, Sinha I, Polin L, Wei WZ, Rathinam C, Flavell R, Massoumi R, Venuprasad K (2011) The E3 ligase itch and deubiquitinase Cyld act together to regulate Tak1 and inflammation. Nat Immunol 12:1176–1183

Index

A. De, *Ubiquitin Chains: Degradation and Beyond*, Springer Theses,
DOI 10.1007/978-3-319-14965-3

MIX
Papier aus verantwortungsvollen Quellen
Paper from responsible sources
FSC® C105338

If you have any concerns about our products,
you can contact us on
ProductSafety@springernature.com

In case Publisher is established outside the EU,
the EU authorized representative is:
Springer Nature Customer Service Center GmbH
Europaplatz 3, 69115 Heidelberg, Germany

Printed by Libri Plureos GmbH
in Hamburg, Germany